Humanitarian Civil Engineering

Institution of Civil Engineers

Publishing

Humanitarian Civil Engineering

Practical solutions for an interdisciplinary approach

Edited by

Georgia Kremmyda

University of Warwick

Published by ICE Publishing, One Great George Street, Westminster, London SW1P 3AA.

Full details of ICE Publishing representatives and distributors can be found at: www.icebookshop.com/bookshop_contact.asp

Other titles by ICE Publishing:

Civil Engineering Special Issue: Humanitarian engineering
Simon Fullalove. ISBN 978-0-7277-6177-4
Civil Engineering Procedure, Eighth edition
Institution of Civil Engineers. ISBN: 978-0-7277-6427-0
Sustainable Infrastructure: Principles into Practice
Charles Ainger and Richard Fenner. ISBN: 978-0-7277-5754-8

www.icebookshop.com

A catalogue record for this book is available from the British Library

ISBN 978-0-7277-6468-3

© Thomas Telford Limited 2021

ICE Publishing is a division of Thomas Telford Ltd, a wholly-owned subsidiary of the Institution of Civil Engineers (ICE).

Cover photo: Workers construct a water drainage and sewer duct in Nampula, Mozambique. Jake Lyell/Alamy Stock Photo

Commissioning Editor: James Hobbs
Development Editor: Melanie Bell
Production Editor: Madhubanti Bhattacharyya
Marketing Specialist: April Nagy

Typeset by The Manila Typesetting Company
Index created by Matthew Gale
Printed and bound in Great Britain by Page Bros, Norwich

Contents

Packing Slip

emerald PUBLISHING

EMERALD PUBLISHING LIMITED

LEEDS, W YORKS FC LS1 4DL
*447580969488
hgraveling@emerald.com
www.emeraldgrouppublishing.com

Page 1 of 1

Invoice #	01497356	Shipment	01497356001	Customer		2065042
Customer PO#	mom3854633	Ship Method	UPS - Ground		Date	12-28-23

SHIP TO:
Baker & Taylor Books
2810 Coliseum Centre Dr
Suite 300
Charlotte NC 28217

BILL TO:

QtyOrd	BackOrd	QtyShip	Product	Description	ISBN	Price Each	Disc	Extended
1		1	9780727764683	HUMANITAIRAN CIVIVL ENGINEERING	9780727764683	59.25		59.25

Subtotal	59.25
Tax	0.00
Shipping	0.00
Total	59.25

RETURN TO:
EMERALD PUBLISHING LIMITED

Comments:

Preface

Humanitarian engineering is the integration of engineering with sciences, social sciences and arts to design, develop or improve tools, approaches, technology and educational practice needed to promote the well-being of populations facing grand challenges. In current times, where various humanitarian crises are happening, such as natural and human-made disasters, energy and sustainability challenges, fast-growing-population cities, pollution and climate change, we find that civil engineers have a pivotal role to play, enabling our society to progress towards sustainable human development.

From water supply to renewable energy provision, from efficient transport systems to digital infrastructure, from constructing resilient cities to the provision of sanitation facilities, engineering knowledge and application underpin the responses needed by us all to pursue a sustainable future. Because the issues of humanitarianism are not just civil engineering problems, there is need to engage with other professions and stakeholders to embrace and exploit combinational expertise, and to introduce approaches and practices that enable sustainable and safe applications.

As demonstrated through the various United Nations (UN) dialogues, notably the Hyogo Framework for Action 2005–2015, the Sendai Framework for Disaster Risk Reduction 2015–2030 and the UN Sustainable Development Goals 2030), there is (a) recognition that building resilience against humanitarian challenges is central to effective global action; (b) lots on explicit and implicit discussion about resilience and transformative change; (c) increased recognition of the importance of humanitarian intervention; (d) a need for promoting developments across different disciplines and sectors to have a global humanitarian impact; (e) and a focus on the advancement of education, learning and training for skills.

Objectives of the book

This book makes the case for the pivotal role of civil engineers in tackling global humanitarian challenges. Enabling the safety of life and property against humanitarian challenges is the highest call of the civil engineering profession. Nevertheless, it cannot be done by civil engineers alone; it requires partnership and collaboration. It requires the introduction of humanitarian engineering for a holistic synthesis to develop sustainable solutions that integrate social, environmental, cultural and economic systems.

Humanitarian engineering can be used as the catalyst for the change that the world needs, crossing knowledge boundaries (within academic disciplines, governments, government internal functions, companies and sectors, and the boundaries between these domains), with interdisciplinary approaches and practices that enable sustainable, safe applications from water supply to renewable energy provision, from efficient transport systems to digital infrastructure, from constructing resilient and safe cities to the provision of sanitation facilities.

The book aligns humanitarian engineering with civil engineering practices, and adds on the work already being done in the developed and developing world, with a focus to serve effectively and responsibly, and primarily, the most vulnerable and unsupported communities in the world. The book aims to identify best practices that are tangible, practical and relevant to ensure safe engineering against global challenges. The book, through a series of case studies, combines concepts and tools traditionally used by practitioners and development agencies with civil engineering practices, project management and systems thinking. When blended, these concepts, frameworks, practices and lessons learnt offer civil engineers better methods to manage the difficulties inherent in community development projects. The level of competence needed from civil engineers to respond to humanitarian needs is also discussed in a context of unique and difficult to measure skillsets.

This book shows the passion of the authors and editor in the application of civil engineering in humanitarian settings. It highlights the most select literature and provides meaningful professional experiences. The book showcases the pivotal role of civil engineering in the solution of global humanitarian challenges, enabling our societies to progress towards a safer future. The book intends to be a coherent and accessible document summarising the knowledge available in humanitarian engineering. It is believed that the book represents a unique and timely compendium on this topic. The book is addressed primarily to engineers – professionals and university students – interested in global challenges and development projects, whether they reside in developed or developing countries. Development workers and practitioners may also find parts of this book useful, especially if they are interested in the technical aspects of humanitarian engineering and sustainable development.

Georgia Kremmyda

About the editor

Georgia Kremmyda

Georgia is a professor, the director of studies in civil and environmental engineering and the deputy head of teaching at the School of Engineering, University of Warwick. Georgia completed her undergraduate, postgraduate taught and doctoral studies at the National Technical University of Athens, Greece. Before joining Warwick University in 2015 and for a period of 15 years, Georgia shared her time between carrying out leading engineering roles in industry and delivering teaching and research in academia. Her teaching and research interests are oriented but not limited to the design of earthquake-resilient structures, sustainable cities, infrastructure for emergencies and the delivery of interdisciplinary, problem-based education, especially in resource-constrained environments.

Within her academic role, Georgia is leading initiatives related to the development of innovative programmes that are promoting equitable access to education, in science and engineering. Georgia leads the Warwick University humanitarian engineering programme and co-leads the Warwick Global Research Priority in Sustainable Cities, both aligned to the United Nations Sustainable Development Agenda 2030. On an international level, Georgia has been elected as Vice President (Conferences) of the International Network of Women Engineers and Scientists for 2021–2023. Within Warwick University, she is the deputy chair of the Institutional Athena SWAN Self-Assessment Team, leading the subgroup working on 'Organisation and culture for inclusivity, diversity and equality' and 'Support to trans people'. She is also the chair of the STEM Education and Student Experience workstream within the framework of the Warwick STEM Grand Challenge.

About the contributors

This book would not have been possible without the incredible and dedicated efforts and contributions of its many authors, who provided us with articles reflecting their knowledge and long-standing experience in the humanitarian and engineering sectors.

Joseph Ashmore

Joseph Ashmore leads the Shelter and Settlements team in the International Organization for Migration (IOM) Department of Operations and Emergencies. The team supports programming that assists between 4 and 6 million people annually. Joseph has over 18 years practical experience with 15 humanitarian organisations in humanitarian shelter and settlements programming. He has also led on Shelter Projects for the development of a range of global consensus-based guidance documents on reconstruction, shelter provision, settlement planning, gender-based violence (GBV) risk reduction and strategic settlement issues.

Simon Bird

Simon Bird is a Chartered Engineer with varied experience of mainstream civil engineering, research, rural development and disaster relief. He is passionate about applying engineering skills in humanitarian settings, and has mentored young engineers and helped with RedR hands-on training. Presently he works part time as a tunnel engineer for Mott MacDonald and also as infrastructure advisor for Medair, a Swiss-based non-governmental organisation (NGO). His most recent assignments for them have been in Lebanon – in providing shelter for Syrian refugees and repair of apartments damaged by the Beirut explosion.

Gustavo Cortés

Gustavo Cortés has a PhD in structural engineering and is a professional engineer (PE) in Texas, USA, and a member of the Swiss Society of Engineers and Architects (SIA). In addition to his experience in the private sector, Gustavo was an associate professor at LeTourneau University in Texas. Presently, he works as the senior shelter and infrastructure advisor for Medair, advising country programmes in Africa, Asia, the Middle-East and Central America in the design and construction of buildings and infrastructure. Gustavo organises regular training workshops for Medair shelter and infrastructure staff world-wide.

Hannah Edmond

Hannah Edmond is a Chartered Principal Civil Engineer working in the water sector at Mott MacDonald. Originally from the UK, Hannah is currently based in New Zealand. She is a RedR UK Affiliate and has a core interest in humanitarian relief, international development, and how private sector skills and experiences can be used to support these roles. Her work includes coordination of RedR UK's Hands on Weekends.

Carrie Eller

Carrie Eller is a civil engineer working for Mott MacDonald's International Development Services, specialising in hydraulic engineering. Previous projects have included irrigation rehabilitation, dams and reservoirs, micro-hydropower and flooding. Carrie has worked on small-scale advisory projects with Water Aid and Concern, and volunteered with Engineers without Borders UK and RedR UK.

Bill Flinn

Bill Flinn is a senior shelter advisor at CARE International UK. He is a qualified architect, and has worked in development and humanitarian relief in five continents, and for many years in domestic construction in the UK. In recent years he has been a leading proponent of supporting self-recovery as an appropriate modality for post-disaster recovery, collaborating on research projects with the Overseas Development Institute, University College London, Oxford Brookes University and the British Geological Survey.

Dan Flower

Dan Flower is a principal and the design director at HKS Architects' London office. Combining creativity and design thinking with a sophisticated approach to material and technical solutions, Dan brings over 15 years of experience to his role as the design lead on major EMEA region projects. He has applied his knowledge of both architectural design and masterplanning to a variety of projects, and is dedicated to delivering innovative, efficient environments. Bringing a lifelong passion for applying design solutions to improve the quality of life for all, he has worked as the architectural advisor for Engineers for Overseas Development (EFOD) for 15 years and is also a key studio leader of Citizen HKS, an HKS initiative providing architectural solutions for NGOs.

Ian Flower

Ian Flower is a Chartered Civil Engineer with over 40 years' experience in the construction industry. Ian has worked for consultancies throughout his career, on major construction projects in the UK and Trinidad, and managed an office in Cardiff for Mott MacDonald for over 20 years. As the chair of the Institution of Civil Engineers (ICE) in Wales group in 2000, he was asked to help to solve the problem of inadequate sanitation in downtown areas of Banjol, The Gambia. Ian challenged a team of six young graduates to design a solution to the problem, raise funds, and visit site for 2 weeks in pairs to hire labour, buy materials and supervise the construction of latrine blocks in downtown areas of the city. EFOD was born. Ian has run EFOD from its inception, initially as a subgroup of ICE Wales, and since 2011 as an independent charitable company. EFOD encourages teams of young members of the construction industry to enhance their experience by working together to deliver projects for the benefit of some of the rural poor in Sub-Saharan Africa. He is a fellow of ICE, and was awarded an OBE in 2006 for services to civil engineering and international development. Together with Helen, his wife, he also runs SaltPeter Trust, a Christian charity that works with Baptist churches in Teso, Uganda, providing educational, medical and agricultural support for widows and orphans.

Step Haiselden

Step Haiselden leads CARE International's Emergency Shelter Team based in London. He is a Chartered Structural Engineer with considerable experience in private practice as well as the charity sector. His humanitarian work in both shelter and water, sanitation and hygiene (WaSH) has taken him to many post-crisis contexts in the Caribbean, Africa, Asia and the South Pacific. He and Bill Flinn first worked together in the aftermath of the 2005 Pakistan earthquake.

Lewis Kelly

Lewis Kelly is a Scottish architect, and technical consultant for the International Federation of Red Cross and Red Crescent Societies (IFRC) Shelter Research Unit. His interests lie in the role of architecture and the built environment addressing the environmental and social challenges faced in society. Building on training from Edinburgh University and further training in sustainable design in Catalunya, he has engaged in field-based research into architecture and urbanism in developing contexts, having worked closely with humanitarian and development organisations. With the IFRC Shelter Research Unit he has researched environmentally informed humanitarian shelter and settlement approaches, and the integration of circular economy approaches into post-disaster reconstruction operations.

Jim Kennedy

Jim Kennedy has a PhD in the design of refugee camps from Delft University of Technology, and has more than 15 years' experience in programming, coordination and guidance development in the humanitarian shelter sector. Jim has worked extensively with the UN, the Red Cross and international NGOs, in both armed-conflict- and natural-disaster-related responses, and is also the author of a number of articles and booklets on various aspects of shelter response, including site planning.

Ellis Lui

Ellis Lui is a civil engineer in the water sector at Mott MacDonald, working on small to large non-infrastructure and infrastructure potable and waste water projects. Ellis is currently deployed in Bangladesh, supervising construction works on a new potable supply scheme. His technical expertise includes pipeline and hydraulic design, civil infrastructure design and construction supervision. Ellis has previously organised disaster relief workshops with RedR UK and delivered NGO-affiliated design schemes at varying professional capacities.

Robert O'Toole

Robert O'Toole is a winner of the National Teaching Fellowship Scheme, and is based in the UK but works internationally. He specialises in developing and implementing transformational strategies that work for real people in schools, colleges, universities and businesses. Research led and practically focused, he uses empathetic and participatory methods to work with communities to create designs that fit, stick, spread and grow – a definition of good design developed in his PhD thesis on design thinking. Collaborating with other colleagues, Robert has developed a suite of interdisciplinary undergraduate and postgraduate for-credit courses at the University of Warwick, developing students as designerly change agents.

Tom Newby

Tom Newby is a Chartered Structural Engineer and a specialist in emergency shelter provision in humanitarian emergencies. He was until 2019 the head of humanitarian for CARE International UK, and prior to that was he was the global emergency shelter lead for CARE International. He has worked in humanitarian responses in Haiti, the Philippines, Nepal, India, Jordan and Lebanon, among others. He has a strong interest in the power dynamics of humanitarianism and engineering, as well as gender equality, inclusion and meaningful accountability in humanitarian responses.

Daniel Paul

Daniel Paul started his professional career in the British military, before moving into project management for the humanitarian sector. He then turned to security risk management, ensuring the safety of those working overseas – an area he is passionate about. He currently works as a security advisor and consultant, as well as lecturing at Coventry University on disaster management and emergency planning. Daniel holds a PhD, during which he studied how organisations keep their staff safe in high-risk environments and how this can be improved with knowledge management. Daniel also has a bachelor's degree in international security and disaster management.

Alberto Piccioli

Trained as an architect/engineer and urban designer, Alberto Piccioli worked on participatory planning and design for community buildings and public spaces before joining the humanitarian shelter sector with the IOM in Geneva. He led the development of two editions of the interagency publication *Shelter Projects*, as well as provided remote and field-level support to IOM shelter and settlement operations and coordination across several countries. Since 2020 he has been based in Maiduguri, focusing on site planning and shelter operations in North East Nigeria.

Regan Potangaroa

Regan Potangaroa is a professor at the School of Architecture at Victoria University of Wellington, New Zealand. His professional background is 25 years as a structural engineer with 14 years as an academic. He has also completed over 200 deployments with the UN, the IFRC, national Red Cross societies and some of the larger international NGOs such as

CARE International. He is an associate trainer with RedR Australia and a delegate with the NZ Red Cross. He has lived and worked in 23 countries. He has master's degrees in engineering, architecture and business management, a bachelor's degree in engineering and a PhD in architectural engineering.

Amina Saoudi
Amina Saoudi has worked in the humanitarian sector for the past 13 years, including with the UN High Commissioner for Refugees, International Committee of the Red Cross, Shelter Centre and IOM. After joining IOM in 2010, Amina worked with the Shelter/Non-Food Items (NFI) and Camp Coordination and Camp Management (CCCM) Clusters in Pakistan, South Sudan and Myanmar. In 2016, Amina joined the Global CCCM Team in IOM's Department of Operations and Emergencies. She currently manages IOM's Global project, aiming at improving operational response to GBV in displacement crises, including in CCCM, Shelter/NFI, Site Planning and Displacement Tracking Matrix operations.

Seema Singh
Seema Singh is a professor of economics at Delhi Technological University, India. Largely interested in issues related to gender, engineering education, employment and labour market, she has published articles in refereed journals and presented papers in national and international seminars and conferences. She has also successfully completed several research projects sponsored by Indian organisations such as the University Grants Commission, the All India Council for Technical Education and Indian Council of Social Science Research, and international organisations as the United Nations Development Programme and the World Health Organization. She is a member of the editorial board and a paper reviewer for many journals. She is a INWES Board Member and the Chair of the nomination committee of the International Network of Women Engineers and Scientists (2021–23) and the president of the University Women Association, Delhi (2019–2022). She has been a joint secretary of the Indian Society of Labour Economics since 2006, and a vice-president of Women in Science & Engineering (WISE) – India since 2011.

Matthew Sisul
Matthew Sisul is a civil engineer, humanitarian engineering practitioner, and adjunct professor, who specialises in risk, planning, design, construction, monitoring, and evaluation in

developing and post-disaster communities. He is an adjunct assistant professor at Columbia University, where he teaches engineering for developing communities, a class where undergraduate engineering students grapple with the ways that engineering design and technology can impact the lives of individuals living in poverty around the world. Topics include the theory of change, logical frameworks, stakeholder analysis, problem identification, project appraisal, risk analysis, impact indicators, and monitoring and evaluation.

Sally Sudworth

Sally Sudworth is a Chartered Engineer and environmentalist currently leading on the Net Zero Carbon strategy for infrastructure at the Environment Agency. Sally is Global Head of Sustainability and Climate Change at consultants Mott MacDonald. Sally was a director for Engineers for Overseas Development, an engineering charity that provides development opportunities for graduates to become professionally qualified. Its community-led projects are based in Sub-Sahara Africa and help to alleviate poverty and improve the quality of people's lives. Sally is a fellow of ICE, the Institute of Asset Management (IAM) and the Women's Engineering Society (WES). She is working on the ICE Carbon Project, and climate change groups at the IAM and WES. Sally has over 30 years' experience of working in design and construction in both the private industry and the public sector, managing multi-million-pound programmes of work for flood risk, asset management, river and canal restoration, and highways.

Antonella Vitale

Antonella Vitale is an architect, urban designer and an established specialist with over 15 years' experience in humanitarian shelter and settlement, emergency and early recovery response programmes. She is a member of professional rosters, including the IFRC Field Assessment and Coordination Team and the IFRC Early Recovery Surge Team, and she has collaborated with numerous international organisations and NGOs, on pre- and post-disaster programmes. Between 2004 and 2011 she was the co-director of Shelter Centre, and between 2018 and 2020 she directed the IFRC Shelter Research Unit. Her current focus is on the integration of sustainable construction and circular economy principles into humanitarian post-disaster operations and disaster risk reduction aiming at mitigating climate-induced migrations, fostering adaptation initiatives and, ultimately, managing retreats.

Douglas White

Douglas White is a senior civil engineer at Mott MacDonald specialising in hydraulic structures. His professional interests encompass canals for irrigation, reservoirs for water resources, and flood control, hydropower and drainage networks. Douglas has undertaken design and construction supervision roles in Uzbekistan, Georgia and Malawi, as well as coordinating humanitarian training workshops on behalf of RedR UK.

Georgia Kremmyda
ISBN 978-0-7277-6468-3
https://doi.org/10.1680/hce.64683.001

Chapter 1

Humanitarian engineering: what, why, where, who?

Abstract

The chapter defines humanitarian engineering and comments on the links between emergency relief, sustainable and international development. More knowledge is needed to increase our understanding of the myriad of drivers of humanitarian challenges. Global awareness needs to be combined with information about the exposure and vulnerability of communities to identify those at most risk. The chapter makes reference to relevant United Nations (UN) dialogues with links to where humanitarian engineering is applicable, why, and who are the main actors. The chapter introduces relevant aspects of local service delivery environments: the actors, agents and stakeholders; the regulatory environment, including applicable standards and codes; the role of central and local governments; and the processes within a typical project cycle. Drawing upon two recent disasters in Haiti, the earthquake of 2010 and Hurricane Matthew of 2016, the chapter explores the role of civil engineers in post-disaster recon-struction contexts and the ways in which those roles and attendant responsibilities may shift from traditional approaches, depending on the project context.

HUMANITARIAN ENGINEERING: THE MESSY REALITY OF DOING GOOD

Tom Newby

Introduction

Every disaster is an opportunity. An opportunity that is frequently seized by the powerful to enrich themselves and entrench their power (Klein, 2007). But also an opportunity that could be used by communities and responders to change things for the better (Twigg, 2015). Good humanitarian engineering projects after disasters have the potential to save lives, to reduce disaster risk, to increase people's resilience, and to address myriad social, economic and environmental problems. But in reality they very often fall short, and sometimes well-meaning projects make things worse. This sub-chapter examines some of the fundamental reasons for this. It draws mainly on examples from the humanitarian shelter sector, but is applicable across all areas of humanitarian engineering.

What is humanitarian engineering?

There's a wide range of engineering that is relevant to humanitarian contexts. The more basic emergency engineering is set out fairly comprehensively in *Engineering in Emergencies*

(Davis and Lambert, 2002). But there is much, much more to engineering in emergencies than the makeshift and make-do approaches set out in it. More advanced engineering becomes highly relevant in urban emergencies and in post-disaster and post-conflict reconstruction, ranging from reinstating urban power and water systems to the construction of multistorey buildings and bridges. From a purely technical point of view, these are conceptually not so different from engineering in non-humanitarian contexts. The difference comes from the urgency and the context, rather than from the engineering problems themselves.

'Humanitarian engineering' is not a recognised engineering specialism. Rather than an area of technical engineering expertise, it is better described as a set of experiences and competencies that allow the recognised engineering specialisms to be successfully applied in humanitarian contexts. These contexts can themselves vary enormously – from projects to support vulnerable people failed by social safety nets in rich countries to those in major complex emergencies affecting whole populations. Many professional engineering institutions' codes of conduct make some reference to engineering in the public interest and for the public good, and to safeguarding the future (ICE, 2004). Engineers should be applying their skills across the board for the long-term good of humanity, and this aligns with a broad definition of humanitarian as *being concerned with or seeking human welfare*. But even if this were truly the case, it would not be sensible to define all engineering as humanitarian. The only consistent way to define 'humanitarian' engineering must be projects and work that is carried out in accordance with humanitarian principles and ethics (Slim, 2015). While humanitarian principles normally refer to humanitarian aid given in response to conflict and disaster, they can easily extend to engineering projects in a wide range of contexts and settings. The foremost of these are

(*a*) Humanity: everyone has a right to receive and offer humanitarian assistance, without strings attached.
(*b*) Impartiality: assistance is given regardless of the race, creed or nationality of the recipients and without adverse distinction of any kind. Assistance priorities are set on the basis of need alone.
(*c*) Neutrality: assistance will not be used to further a particular (party) political or religious standpoint.
(*d*) Independence: assistance will not be used as an instrument of government foreign policy, or indeed as an instrument of any donor's priorities.

Humanitarian principles also include respect for culture and customs; build on local capacities, participation, do no harm, build resilience (or 'build back better' – see below), accountability and dignity. Others have defined all of these in great detail, and they are all described in the Red Cross Code of Conduct. Any engineer or engineering organisation seeking to undertake humanitarian engineering must, implicitly, understand and apply these principles, and insist that they are applied throughout the entire project. This goes beyond an understanding of the simple definitions, to knowing how to use and apply the principles in practice; knowing which principles are more important when they conflict with each other and how to balance them. The first two principles are most important – and no project or undertaking that does not comply with them can be described as humanitarian. Apart from the first two, the principles are not a rigid set of rules, and can conflict with each other. Rather, they are a set of essential tools that help navigate the messy realities of humanitarian contexts.

So while it would be true to say that engineering that only complies with the first two humanitarian principles is 'humanitarian' engineering, that would not be good humanitarian engineering. It would likely be harmful to some or all of the stakeholders involved. Good humanitarian engineering must take due consideration of all the humanitarian principles, and carefully and thoughtfully weigh them in the design and implementation of projects.

Implicitly, that means engineering that aims to do good is not, inherently, humanitarian engineering. The bar is set considerably higher, and good intentions are not enough.

What is humanitarian engineering for?

The previous section describes what humanitarian engineering is, but what is it for? At its most basic it must be to impartially preserve human life, and human dignity, in the face of shocks and challenges. And that is where the engineering set out in engineering in emergencies comes into its own. But that cannot be the extent of it. There are recurring debates about whether that is enough, and the potential harm of short-term interventions with no consideration of the longer term. Are humanitarian interventions purely to save lives and preserve dignity, or should they also be supporting recovery and reconstruction, or even addressing some of the root causes of the vulnerability and injustice that caused the particular disaster in the first place?

If do no harm, accountability, resilience building and building on local capacities are important humanitarian principles, then it is inevitable that the longer term must be considered in all humanitarian projects. And engineering projects tend to involve, by their very nature, long-term interventions. Apart from the most flimsy of installations, engineering projects provide infrastructure and structures that are both costly (financially and environmentally) and durable. They will likely be around for a long time. It is rarely acceptable to say that an engineering intervention must consider only immediate life-saving needs and nothing else.

Furthermore, after conflicts, emergencies and disasters around the world there are persistent calls to 'build back better', and statements of 'never again'. From the 2004 tsunami to Grenfell Tower, the expectations of post-emergency responses are that they avoid anything like this being possible in the future. Given that all disasters are more the result of injustice and human decisions than they are of chance (Chmutina *et al.*, 2019), the expectations and hopes of survivors, save perhaps from the privileged few, are never that things be put back the way they were before.

So we come to 'build back better' – an alluring phrase to engineers that became increasingly widespread after the 2004 tsunami and the 2010 Haiti earthquake, and which seemed to encapsulate in three words exactly what humanitarian engineering should be all about. It was, in 2015, adopted in the Sendai Framework for Disaster Risk Reduction (UN, 2015), and the concept has been further developed by the Global Facility for Disaster Reduction and Recovery (GFDRR, 2017) and the UN Office for Disaster Risk Reduction (UNISDR, 2017) to be about increasing the resilience of the built environment and infrastructure. However, in general usage it often has a much broader meaning, in addition to stronger infrastructure, encompassing better rights for poor people, more just systems of governance and much more. And, indeed, it is unrealistic to think more resilience can be achieved without also addressing such issues. But it is telling that the most vulnerable people, who should be those prioritised in humanitarian interventions, are rarely asked what 'better' would look like. Instead, what is better is decided by remote and powerful people, in government, in NGOs and in multilateral agencies.

While humanitarian interventions are almost never as fleeting and transient as may be hoped, they are equally too hurried and simplistic to actually solve many of the complex societal, economic, environmental or other injustices facing people in need. A well-meaning desire to use the window of opportunity presented by disaster to build back better all too often results in exactly the opposite for the most vulnerable people who should be those prioritised in humanitarian interventions. Without agreement from them on what actually is better, humanitarian projects are doomed to fail them (Crawford, 2018).

Pragmatically, there is a balance to be struck between serving people's immediate and urgent needs and addressing the root causes of people's exclusion and vulnerability. The humanitarian sector has for many years been grappling with how to do this, running through repeated debates on the 'humanitarian–development continuum', 'linking relief and development' (Buchanan-Smith and Maxwell, 1994) and, most recently, the 'humanitarian–development–peace nexus'. There is no simple answer to what humanitarian engineering is for, but to get the balance right between short-term needs and long-term goals, a humanitarian engineer needs to move well out of the comfort zone of most engineers and designers, to consider issues of lived reality, identity, colonialism, racism and patriarchy. Humanitarian engineers must consider very carefully what their role is, and cede a considerable level of control over what is designed and built to those they seek to help (Institution of Structural Engineers, 2020). Engineering expertise is an essential part of good humanitarian response, of effective and equitable recovery from disasters, and of just long-term development, but justice and equity do not happen as automatic side-effects of technical excellence. In fact, most engineering projects are unequal and unjust by design (Criado Perez, 2019). Across the world, in countries rich and poor, women, girls, people with disabilities, ethnic minorities and marginalised groups are frequently and systematically excluded from all aspects of engineering design and construction. This exclusion is only increased in times of emergency. Humanitarian engineers must take responsibility not just for the technical aspects of engineering but also ensure that what they do is inclusive, equitable, non-patriarchal (Cordero-Scales *et al.*, 2016), anti-racist and decolonial. That is no small challenge.

How to do good humanitarian engineering

Good, technical engineering looks very similar wherever it is. The technical interventions should be appropriate (Schumacher, 1973) to the context. But in most humanitarian situations the framework in which engineering takes place is very different to most engineers' experience of professional practice. The population may be traumatised, oppressed or displaced. Civil society may be very limited in scope or capacity. The government may be weakened, fragmented or oppressive. Resources may be constrained and the operating environment very far from enabling. There will be a degree of chaos. Humanitarian engineering takes place in a very messy reality.

In highly specialised, professional settings it is not hard for engineers to insulate themselves from the ethical, cultural and socio-economic aspects of their projects. Things such as community participation, economic analysis, social impact, environmental impact and ethics are dealt with by other people. Clients, other specialists, lawyers and regulators are responsible for those, and engineers are simply responsible for the successful delivery of the brief. Some may question the brief from a technical point of view, and propose different approaches. Few will question the premise of the project, the process that has been used to land upon it, or intended and unintended outcomes of the work.

This isolation from the real-world impact of engineering is an abdication of professional responsibility in any case, but in humanitarian contexts it is entirely untenable. All those specialists who limit the potential harm of projects, and all those lawyers and regulators who ensure projects and organisations comply with their responsibilities, either do not exist or are totally overwhelmed in most humanitarian contexts. In humanitarian contexts it is impossible as an engineer to pretend your project exists in a neutral, apolitical space. As an engineer you have to engage fully with the messy reality in which your projects are to be implemented. The job of implementing an engineering project becomes one of navigating this messy reality.

The first thing to understand is that this messy reality looks remarkably different to all the different stakeholders who will be involved in or affected by any engineering project.

Disasters most severely affect those in society who are most vulnerable, most excluded and most marginalised. The effect of disasters is rooted in injustice and inequality. The quickest and most consistent aid is almost always provided by communities themselves, by neighbours, by family, by local authorities, through communal efforts and cooperation. But, in general, aid is given by those with more wealth and more power, and received by those with less wealth of power. The fairness, justice and equity of the process of providing aid can look very different viewed from the perspective of the aid-giver and the aid-receiver.

Engineers undertaking humanitarian work must understand the historical and present context of patriarchy and colonialism in which they are working. They must understand that their own identity and position, and the identities and positions of the people they work with and for, will significantly affect what they are doing in many ways. The effect of their identity on how they will be perceived and treated by others is likely to be considerable, and their behaviour and actions will, in turn, affect others who do not have such immensely privileged positions. A young, White, professional engineer and 'ex-pat' aid worker with the confidence of an expensive education will often have their opinions valued considerably more than a local Muslim female engineer and member of the local staff team – even if said female engineer has decades of relevant experience and knows much more about the subject. A young, unmarried woman from a marginalised community may be unwilling and indeed unable to speak to an older, male, senior engineering advisor even if she has information of great relevance and importance to impart. Humanitarian engineers must understand the power imbalances inherent in the work they are doing, and consistently and proactively adjust their approaches to address them.

Intersectionality

Kimberlé Crenshaw (1989, 1991) coined the term 'intersectionality' in 1989, as the effect of the intersection of race and sex on the discrimination faced by Black women (Smith, 2013) – something already recognised in the earlier work of Black feminists such as Sojourner Truth, Mary Church Terrell, Nannie Burroughs, Fannie Barrier Williams, Anna J. Cooper (1892) and many others (Guy-Sheftall, 1995). The term came into common usage in the 2000s, and is now widely used as the broader intersection of all identities and the particular discrimination and disadvantage that results (McCall, 2005).

In summary, all people have a set of differing identities that combine to make them who they are. These include nationality, race, gender, sexual orientation, education, class and many more. Different identities result in differing levels of privilege and power, or discrimination and exclusion. These different identities overlap and intersect, and those with multiple dis-criminated-against identities suffer increased disadvantage and marginalisation that manifest in ways particular to the intersecting identities in question. Understanding intersectional identities and discrimination is important in dismantling the injustice and inequity so many people face.

International aid, ranging from voluntourism to professional aid work, is built on foundations of patriarchy and colonialism. Colonising countries systematically extracted wealth and resources from colonised countries, and oppressed and murdered their people. Much of the inequality and injustice that exist in the world today, which aid should be seeking to address, results from over 400 years of colonisation. Colonial practices persist through global financial systems and taxation, limits on the movement and migration of people, ownership of land and property, and many other routes.

The fact that wealthy, predominantly White, people from Europe and North America have the wealth, power and privilege to deliver aid to predominantly Black and Asian people *elsewhere*, who do not have such wealth, power and privilege, is a colonial artefact. The fact that elites within some countries have the wealth, power and privilege to deliver aid to the poor and marginalised is a result of patriarchy and often colonialism. Given this, and in the absence of more effective remedy, aid should not be seen as charity, as magnanimity, but rather as rec-ompense. Aid is owed. In that context, the aid-givers are not those who should be deciding what aid should look like or how it is spent and delivered. Aid-givers should use their wealth, power and privilege according to the wishes and choices of those who have been systemati-cally disadvantaged because of their wealth, power and privilege.

For humanitarian engineering, this means that everything from the project conception to the final product must be according to the vision and wishes of the end-users and those impacted by the project, and not according to the vision and wishes of the engineer, their employer or their funder. People struck by disaster, on top of systemic injustice and poverty, face a range of day-to-day risks that most professional engineers, who are almost universally wealthy and privileged, cannot even imagine (Kennedy and Newby, 2018). The engineer, whether local or international, must not impose their viewpoint and lived experience on others, particularly their viewpoint of what will or will not make people safer or will make their lives better.

What happens if they do looks something like the following. After many disasters, large numbers of houses are damaged and destroyed. This always results in a multitude of calls to 'build back better'. Architects and engineers, supported by institutional donors and powerful politicians, descend on the damaged communities with grand plans and designs for better houses. 'Never again shall your houses fall down', people are told. People agree this is a good idea, and this is taken for comprehensive approval without any further discussion. Identikit 'bombproof' structures spring up, bearing little resemblance to what people might themselves consider to be a 'better' house. Important spaces such as prayer rooms, or kitchens, are not included because the budget has been spent on strength. Key design features such as a veranda, or a loft for storing grain, are omitted, because they do not contribute to safety. To make them better, houses are moved to safer sites, far from livelihoods, infrastructure, services and community support. Recipients do not complain very much, because they are not asked, and they do not have the power to raise objections. They are afraid they will be left with nothing if they dissent or offend the engineer. What results is a lot of money spent on houses that, although safe from easily defined once-in-50-year hazards, do not meet any of their occupants' actual needs. People are safe from storms and earthquakes, but they no longer have a livelihood and their children cannot get to school. Over time the house crumbles, as the occupants do not have the resources to maintain it. The burden of all of this falls disproportionately on women (Bradshaw and Fordham, 2013). The promised better house really is not better, and has left people more marginalised and less safe. The architects, engineers, donors and politicians go home with pride at the end of construction, having successfully protected people from the only risk they understand.

A further consideration for humanitarian engineers is whether their work builds or undermines the engineering capacity where they are working. Although the strength and prestige of the engineering professions vary, every country has its own engineering professionals who will remain long after any international responders have left. If engineering and engineers play a vital role in the resilience of society and of infrastructure, then the strength of the engineering professions is a key indicator of that resilience. Projects that side line or undermine local engineering capacity and institutions are harmful. Engineering projects that fail to build the strength and capacity of the engineering professions have missed an opportunity to decolonise engineering and build true resilience (or build back better, if you will).

These are what patriarchal humanitarian engineering projects look like. They may be well meant, but are inherently self-serving, and ultimately harmful. Whether it is entirely abandoned villages built after the 2001 Gujarat earthquake (Sanderson and Sharma, 2008), communal water supply systems falling into disrepair, or toilets being used as plant pots (CARE India, 2016), the history of humanitarian aid is littered with well-intentioned technical solutions that did not align with the complex human realities of their intended users. It is sadly far from a rarity.

A key factor in avoiding the scenarios above is true and meaningful accountability to disaster-affected people. An engineer undertaking humanitarian engineering must understand who they are accountable to, and who is responsible for defining the objectives of a project. In most professional engineering contexts, an engineer responds to a brief set by a client. That client has worked out what they need and what they want, and the engineer is accountable to the client for providing that. If the client gets the brief wrong, a helpful engineer may point that out, but ultimately it is the client's own responsibility. If the engineer does not comply with the brief, they have to fix it, or they will not be employed again, and might even be sued or prosecuted.

In humanitarian settings it is often very unclear who the client is, and what the brief is. Who is a humanitarian engineer really working for and who are they accountable to? Is it the donor or donors who are paying for the work? Is it the organisation undertaking the response? Is it the end-users, the affected people? It is, in reality, most often the donor. It should, of course, be the end-users. The very name 'humanitarian response' implicitly requires the work to be for the benefit of humanity, of people. In reality, the affected people rarely have any involvement in defining the project, setting the brief or in evaluating the project's results. They do not have the option of firing or not employing the engineer again. They almost never have recourse to legal remedy. Yet they are the ones who have to live the consequences.

This lack of true accountability is both ethically unacceptable and a fundamental obstacle to the meaningful success of engineering projects. Regardless of the particular set-up of the project, a humanitarian engineer will take responsibility for ensuring genuine participation of affected people in the definition and delivery of the project, and will make sure that there are safe and effective ways for people to make suggestions, give feedback and make complaints – and, vitally, that these are acted upon. Ultimately, the engineer should treat affected people as they would clients in more conventional professional practice.

If engineering in humanitarian contexts is to be successful, it is first necessary to come to an agreement with the users or recipients of the engineering in question about what success looks like. While engineers are likely to see success purely in terms of technical quality, users of their work are likely to have an entirely different view of it. As noted in a briefing on research findings from multiple agencies' shelter responses after super typhoon Haiyan in the Philippines (Zarins *et al.*, 2018)

> A telling failing from years of shelter programming is that the sector itself does not have a common understanding of, and even worse we don't tend to agree with our project beneficiaries on, what constitutes success. Too often our only measures of success are numbers of shelters built, and sometimes the occupancy rate should we have the opportunity to go back and look at a later date. Given that this research shows that affected people usually contribute more resources and more value than the external shelter actors do, the lack of control they have over what assistance projects are trying to achieve is particularly shocking, and places the power imbalances we perpetuate into stark contrast.

Having agreed on what actually needs to be achieved, it must be recognised that not every humanitarian outcome requires engineering or infrastructure solutions. Humanitarian engineers must be willing and able to envisage non-technical interventions. Different options should always be explored with the affected people, to confirm that engineering really is what they lack. For example, in the case of a disaster resulting in destruction of houses, it might actually be that the affected people know how to build something, but they just do not have the money. It may be that they just need land tenure or legal recognition, and then they will be able to borrow or invest their own money. It may be that they can rent somewhere if only their livelihoods could resume, and that needs an entirely different intervention. Perhaps engineering is needed, but it is not enough, because even with the engineering input some people will be prevented from rebuilding because of local politics or exclusion. If these conversations are not held, and the questions are not asked, humanitarian engineers travel the world building poorly thought-out and unnecessary infrastructure.

In many disasters and emergencies, engineering really is a key component of success. When that is the case, humanitarian engineering must find the best possible balance between ceding control over projects to the people for whom they are being done, and maintaining sufficient technical oversight and rigour to ensure the engineering is appropriately robust and safe. Where the right balance lies will vary enormously. It is often not appropriate for that balance to be determined by codified standards written for wealthy countries where the landscape of risks faced by the population is entirely different. A process that allows disaster-affected people to make informed choices about the cost and robustness of engineering done on their behalf, versus their other priorities, is essential. Such decisions cannot be made solely by the engineer, with their particular viewpoint of the risks people face. Neither should people be left to make those decisions alone, without support and relevant guidance from engineers or other experts.

There are many ways to cede power and control in aid work, and in humanitarian engineering projects. One of the simplest is to give control of the budget to the people in whose name the project is being done. Giving people unrestricted, unconditional cash has been shown to lead to better outcomes in a range of sectors (Bailey and Harvey, 2017; ODI, 2015). While care must clearly be taken in how to extend this approach of giving cash to individual households to the construction of infrastructure or other engineering projects, it is impossible to argue that refusing to give control of aid money to the people it is meant to serve is not patriarchal and ultimately harmful. There are many ways to enable community participation in projects (ALNAP, 2009), and to go further and establish participatory budgeting (Cabannes, 2014), community organisation and control of projects and assets (White *et al.*, 2018); such approaches must be central to humanitarian engineering.

When working at the household level (e.g. when constructing individual household shelters or housing), there is little reason to not give the money to the household, without restriction, and to allow them to choose how to spend it. In this scenario the role of a humanitarian engineer becomes one of a consultant and advisor to the household, giving them the information they need to make good decisions about how to spend the money. The role of aid workers becomes one not of giving people stuff, but of helping people overcome the remaining obstacles they face, whether they be ones of regulation, exclusion, discrimination, knowledge or skills – just as it would be in private professional practice in a non-humanitarian setting. The true challenge lies in how to do this at large scale.

There are no perfect humanitarian engineering projects. Disasters and their aftermath are too chaotic, too difficult, too messy, for perfection. But whether it is the locally driven reconstruction of thousands of houses in Pakistan after the Kashmir earthquake of 2005 (Mughal *et al.*, 2015) or the thoughtful and carefully evaluated approach of using community-led total sanitation (CLTS) and participatory hygiene and sanitation transformation (PHAST) approaches in combination by national Red Cross Societies in East Africa (IFRC, 2018a), there are many examples of good projects, both small and large scale, where humanitarian engineering interventions have saved lives, improved lives and, indeed, built back better. When people are in need, it is right that those who are able provide help. But it requires serious consideration, expertise and humility to ensure the help offered is helpful and not harmful.

Conclusions

Humanitarian engineering may be done for a range of different reasons, but the objectives of any particular project must be well considered and realistic, and not in conflict with the humanitarian principles.

Humanitarian engineering is more about the process than the product. The product is the easy part. Humanitarian engineers can only do the good they set out to do if they are able to work in collaboration with, and subsidiary to, the people they want to help. They cannot concentrate only on the engineering product. They need to develop understanding and a set of skills not taught in engineering courses or used in normal engineering practice.

The possibilities of humanitarian engineering to save lives and make things better seem endless. But the messy realities in which they are implemented means they rarely if ever live up to those ambitious expectations. A great deal of humility, thought and realism is needed when embarking on humanitarian engineering projects, whether those are in emergency contexts as for most of the situations described in this sub-chapter, or in more stable 'development' contexts. Unless engineers responding to human need fundamentally revise their view of what the role of an engineer is, they are likely to reinforce the patriarchal, colonial, and unjust systems and structures that lead to disasters and human suffering in the first place. And perhaps such changes in approach could come to be of great value in more conventional engineering too.

ENGINEERING SERVICE DELIVERY IN HUMANITARIAN CONTEXTS

Matthew Sisul

Introduction

Civil engineers working in developing country and post-disaster contexts require a deep understanding of both civil engineering professional practice and the unique challenges of humanitarian engineering interventions. Civil engineers are not only required to apply engineering design best practices of a local context, they must also understand the particulars of the local service delivery environment in which they are working, including the local regulatory environment, gaps in local capacity, and how the realities of a given project may diverge from the familiar norms of projects in the global north.

The next section unpacks what it means to be a humanitarian civil engineer, including issues related to many different hats they are frequently called on to wear, introduces key terminology and defines the audience for this sub-chapter. The third section follows a discussion of the ways in which the development sector in general is shaping the debate around local participation in projects and how that impacts humanitarian civil engineers working in this context. The section provides an overview of key project facets, including service delivery, project teams, project processes, supply chains, and the regulatory and contextual environment. These facets are the lenses through which humanitarian infrastructure projects can be analysed for opportunities to reshape the projects to emphasise best practices from the global north, local participation, and capacity building at the local level. The fourth section provides a brief overview of three case studies to demonstrate how a civil engineer could compare and contrast projects with similar objectives according to the diagnostic framework and draw insight into alternative project approaches. Finally, the fifth section summarises lessons learnt based on the examples provided, insight from the diagnostic framework process, and recommendations for civil engineers engaged in humanitarian infrastructure projects or those that hope to become so engaged in the future.

Level setting

As stated above, engineers working in developing country and post-disaster contexts must not only have expertise in civil engineering professional practice, they must also be able to nimbly navigate the unique challenges that inevitably arise in humanitarian engineering infrastructure interventions. They must bring engineering design best practices to infrastructure in the local context, and they must be prepared to, for better or worse, take on roles that are typically unfamiliar to civil engineers working in the global north.

Of course, even those roles typically assumed by the civil engineer will vary within a particular project and/or programme, just as they do in the global north. Regardless of the role, be it in planning, design, construction or elsewhere in the project lifecycle, civil engineers working within the humanitarian context require an understanding of the unique challenges, both technical and non-technical, that they are likely to face when executing projects in developing countries. While the civil engineering profession has come a long way in developing codes, standards and best practices for projects in the global north, their application in humanitarian contexts is not necessarily a direct translation. Further, non-technical barriers are equally likely to derail a project as those that are technical in nature.

While many of the barriers of humanitarian infrastructure projects are common to most or all such projects, others will vary based on the unique circumstances of the engagement. Humanitarian infrastructure projects are context specific, meaning that the social, environmental, political and historical circumstances of a given location factor heavily into the basis for how the project came about in the first place. The communities targeted by these projects, be they rural or urban, developing, post-disaster or post-conflict, are characterised by an established or perceived need or deficit. The infrastructure delivered may functionally address, for example, the need for or deficit of lack of access to potable water or lack of adequate shelter. The need, however, is a symptom of a larger disfunction, one characterised by the scarcity of resources. In a root cause analysis, the lack of potable water within a community is not caused by the lack of a water system, but, rather, is caused by a lack of some combination of funds, an identified source, treatment, storage, conveyance, distribution and/or technical knowledge to tie all these together; that is, it is a lack of a system for delivery services at the local level that would provide such a water system. Thus, the civil engineer will need to assess the specific deficits of the community in question and develop ways to overcome the circumstances that led to the deficit in the first place.

It should be noted that this sub-chapter is specifically aimed at the large number of external humanitarian civil engineers who are brought into developing countries to fill a gap and support the provision or delivery of infrastructure projects. They are often brought in by funders or programme managers to fill a role that local individuals cannot perform due to limited local technical capacity or to meet project requirements such as specific work experience or certification. While external agents providing humanitarian aid in local contexts is a familiar scenario, this should in no way diminish the efforts of nationals or those from peer countries within the global south working to rebuild their own countries, nor does it imply that nationals cannot be humanitarian engineers. In Haiti, following the 2010 earthquake, numerous Haitian engineers participated in relief and reconstruction via non-governmental organisations (NGOs), government agencies, UN agencies and local businesses, as well as diaspora professional organisations, and certainly Haitian engineers outnumbered expat engineers. The framework set forth in this sub-chapter, however, is a tool to help those who may be unfamiliar with the challenges commonly associated with applying best practices from the global north in developing countries, which are more likely to surface in the context of humanitarian engineers coming from the global north. Though national engineers are not the primary audience of this sub-chapter, they may nonetheless find aspects of it useful for their work.

With this conception of the humanitarian engineering as outsider in mind, the civil engineer is expected to be both an expert in civil engineer practice as well as a development practitioner. While no civil engineer is expected to master all aspects of the various civil engineering disciplines, nor to be intimately familiar with all aspects of the project delivery process, civil engineers in humanitarian contexts will often be asked to provide services outside their area of expertise. Again, this is a symptom of the context – formal service delivery processes may not exist, or they may exist in a form very different from those found in the global north. As a result, familiarity with the business of civil engineering, or an understanding as to how projects are delivered, is an invaluable part of the humanitarian civil engineer's toolkit. On the humanitarian side, from a technical perspective, the civil engineer should be comfortable translating civil engineering practice into the local context, or, better yet, be familiar with best practice with respect to sustainable design and appropriate technology. In terms of development soft skills, the humanitarian civil engineer should be familiar with the shared language of the development community, including best practices, tools and approaches, such as the theory of change, logical frameworks and root cause analysis.

From the perspective of small-scale infrastructure in predominately rural areas, the practice of a humanitarian civil engineer may proceed from planning to design to construction in an informal manner, in which the civil engineer is intimately involved in each step of the process and, possibly, taking on roles that would typically be assigned to professionals from different industries. In larger organised projects with professional funders such as bilateral or multilateral institutions (e.g. the US Agency for International Development (USAID) or the UN, respectively), civil engineering roles may more closely resemble practice in the global north. However, while these projects may require civil engineers with the technical expertise and experience to support projects or programmes in a variety of sectors, throughout the project delivery cycle (planning, design, construction, management, installation, operation and maintenance), civil engineers may also be tasked to provide education and training, technical capacity building, or professional guidance to local engineering and construction industries.

As a result, for better or worse, the role of the humanitarian civil engineers on a given project may expand beyond the boundaries of a typical project in the global north, to include roles typically taken on by funders, developers, planners, architects, maintenance staff and operators. The cause of this blurring of boundaries is generally the same one that drives the inclusion of humanitarian civil engineers in the first place – the project is informal and atypical, in a resource-poor environment where local infrastructure service delivery faces significant gaps or shortfalls. That is, there may not be anyone else around to perform the required roles to execute the project. However, it is for this very reason that humanitarian civil engineers should remain critical of their own role in projects, analysing each project context and diagnosing gaps in the service delivery environment, and, subsequently, recommending alternative approaches that mirror practice in the global north, while strengthening local service delivery and emphasising local participation in all phases of the project process.

Diagnostic framework

Development practitioners have long pushed international organisations to support local systems. In 1990, the World Bank noted that 'major lenders have promoted decentralisation as a means of breaking the power of central ministries, increasing revenue generation and shifting the burden of service delivery onto local stakeholders' (World Bank, 1990). Others have argued that 'NGOs should "also seek to build up the capacity of the state as an integral part of this localised, grassroots work", rather than creating parallel or alternative welfare delivery systems outside the state' (Mohan and Stokke, 2000).

The reality today is that 'localisation' is now part of the development lexicon, and, although not universal, it is an important facet of how outsiders should conceptualise aid (IFRC, 2018b). This shift impacts humanitarian engineers directly, through initiatives such as the Grand Bargain, a 2016 agreement between donors and humanitarian organisations in which the signatories committed to prioritising support and funding for local and national partners in post-disaster response (World Humanitarian Summit, 2016). It is now up to humanitarian civil engineers to understand how to internalise localisation and apply it to infrastructure projects.

Despite the general push for localisation, the implementation of humanitarian engineering projects administered by outside actors such as not-for-profit organisations (both small and large) and bilateral or multilateral institutions often ignores the local service delivery environment. Additionally, projects are conceived, designed and executed in a manner that is at odds with service delivery frameworks common in the global north. A common service delivery framework consists of policy-makers (who set policies and regulations, e.g. building codes, and allocate resources), providers and beneficiaries connected through a series of accountability relationships (Figure 1.1).

Figure 1.1 Model of a simplified generic service delivery relationship

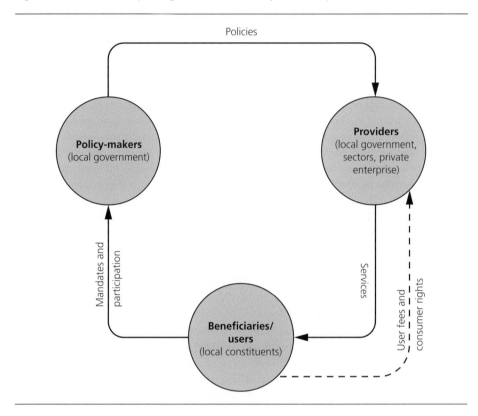

In a functioning ecosystem, a service model for potable water, for example, would consist of policy-makers with mandates to ensure access to water who set regulations on water quality, the quantity of water to be provided and prices, and who select and negotiate agreements with providers to supply water to the beneficiaries. The beneficiaries purchase water from the provider at a tariff rate set by policy-makers. Each entity in the relationship possesses means through which to hold the others accountable – for example, beneficiaries can hold policy-makers accountable through elections in democratic societies, and can hold providers accountable by refusing to pay for services.

Although these approaches embody arrangements that should promote effective service delivery, local reality is often quite different. Local capacity may still be in the process of being built; local planning systems may be in their initial stages or may not exist at all; local governments may not be autonomous and instead take direction from higher levels of government; and non-local non-governmental actors may be extensively employed in service provision, effectively gap-filling for a weak state or local civil society (Marcussen, 1996; Whaites, 1998). In these instances, outside actors that are active locally play an important role in service delivery, and they may do so in a way that bypasses local structures and processes (Figure 1.2),

Figure 1.2 Direct bypass to a generic model for service delivery

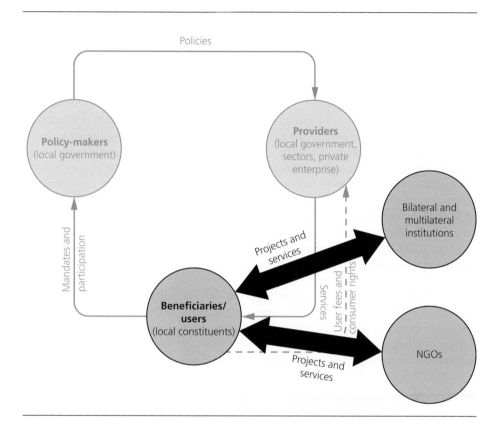

collaborating with local actors in an ad hoc manner or only when necessary. Such behaviour potentially reinforces upward accountability and insufficiently promotes downward accountability (Krishna, 2003).

While bypassing weak or ineffective local delivery systems may seem justifiable in the short term to advance a project, it can have destabilising effects in the long term. Beneficiaries may elect to regularly seek out outside help to solve local problems, and, aggregated over a nation as a whole, bypassing local systems undermines the development of a robust national service delivery environment.

Civil engineers working in humanitarian contexts should be aware of the importance of localisation in the execution of development projects and seek to determine how they can carry out their work in a way that builds up the local service delivery system. For the civil engineer, a simple first step when considering how to strengthen the local service delivery system is to compare the humanitarian project to the typical practice of project delivery in the global north. Infrastructure projects in the global north typically have clearly defined roles for the owner/

funder, design consultant and contractor. For example, like the service delivery model introduced above, the design–bid–build model (Figure 1.3 left), commonly employed in the global north, highlights clearly defined accountability relationships between parties. In the development context, however, the funder is often not the end user of the project, and, instead, a beneficiary group is the recipient of the project and is expected to operate and maintain it upon completion (Figure 1.3 right).

From a project delivery standpoint, it would make sense for the beneficiaries receiving the project to act in concert with the owner, since beneficiaries are often asked to perform roles such as that of the owner once the project is operational. A similar thought exercise can be performed with respect to the contractor and design consultants, although it is worth noting that rather than tasking the beneficiaries with designing and constructing the project, it might make sense to also include local design professionals and private enterprises. The exercise of contemplating enhanced roles for local partners and beneficiaries should be repeated at each stage of the project process, including planning, design, construction, operation and maintenance, and, where possible, monitoring and evaluation (Figure 1.4). In addition, the construction supply chain, the regulatory environment and the project context can be mined for opportunities for enhanced local participation.

Figure 1.5 represents a diagnostic framework that can aid humanitarian civil engineers in brainstorming and mapping the various aspects of the project to the questions they should be asking when considering the level to which the project is truly integrated into the local service delivery environment. Common strategic analysis tools such as SWOT (strengths, weaknesses, opportunities and threats) analysis, stakeholder power/interest matrices and

Figure 1.3 Design–bid–build project delivery model

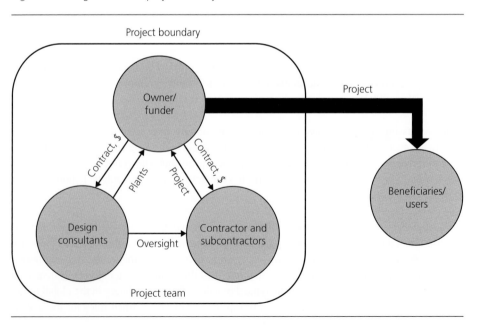

Figure 1.4 Typical project delivery process

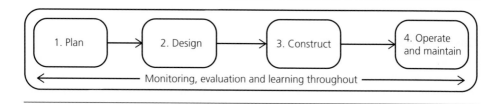

Figure 1.5 Diagnostic framework

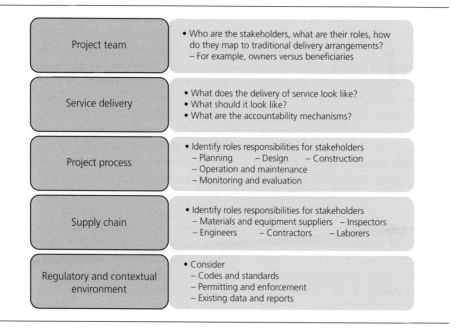

logical framework analysis can be employed at each stage of the diagnostic to identify ways in which to integrate local participants.

The purpose of the exercise illustrated in Figure 1.5 is to interrogate the various aspects of the project and determine alternative approaches defined by sustainability, localisation, participation and empowerment. In the next section, case studies from post-disaster humanitarian civil engineering projects in Haiti are discussed through the lens of the diagnostic framework.

Emphasising local participation throughout the process can improve project outcomes and build local capacity, thereby improving the long-term sustainability of projects. Often, however, a humanitarian civil engineer's engagement in a project has a narrowly defined scope, for example, they may have been brought in only for the purpose of design or construction. Further, in many project environments, the potential for local participation may be limited due to lack of technical capacity. It is therefore incumbent upon the civil engineer to factor in both their own influence and the realities of the project environment before making any recommendations for alternative approaches to project delivery.

Case studies from Haiti

The three case studies arise out of two recent natural disasters in Haiti that led to large internationally funded humanitarian responses. The first two case studies are of projects that were implemented in the aftermath of the 2010 earthquake outside of Port au Prince. The third case study is of a programme that was implemented after Hurricane Matthew struck Haiti in 2016. In both disasters, lives were lost, homes were destroyed, survivors were displaced, and the general economy ground to a halt.

The case studies were projects designed to provide permanent housing for displaced people. Interventions for the provision of housing start shortly after the event and expand after the immediate relief provided (food, water and medicine). Shelter interventions start with emergency shelter (tents and makeshift shelter in camps or on the street). Transitional shelter designed to last 3–5 years is then deployed, followed by large-scale efforts to provide permanent housing over the long term. The sequence of the provision of post-disaster services has a certain logic to it and is suited for top-down, centrally planned interventions. However, there are obvious limitations to this model. In particular, these interventions are planned in a resource-constrained environment: money, personnel and materials are all of limited availability, so the number of beneficiaries decreases as the interventions advance from emergency shelter to transitional shelter and, finally, to permanent shelter. Transitional shelters and camps become permanent settlements with sub-standard infrastructure, creating large, vulnerable populations.

The first case study covers initiatives to provide permanent housing, provisioned by large-scale bilateral and multilateral institutions in the aftermath of the event, on land that was otherwise unoccupied. Specifically, USAID undertook multiple greenfield housing projects in the aftermath of the earthquake in the outskirts of Port au Prince and in the north of the country (US Office of Inspector General, 2014; USAID, 2015). These projects closely mirrored project delivery in the global north. USAID conceived projects in consultation with local government counterparts and then bid out to programme management organisations, who in turn identified contractors and design consultants in the private sector. Local firms were typically not capable of competing with US firms for key roles, and, as a result, US firms generally won the bulk of the work (Macdonald, 2020). Housing was designed according to US standards, and the design consultants had leeway in selecting construction materials. Project sizes varied. In Cabaret, outside Port au Prince, USAID funded the construction of a housing development of 156 units, and in Caracol-EKAM, outside Cap Hatien, USAID funded 750 units (US Office of Inspector General, 2014).

The second case study involves work performed by Build Change, a non-profit organisation based in Denver, Colorado, that specialises in the construction of disaster-resistant buildings and training of local professionals. In response to the 2010 earthquake in Port au Prince, Build Change focused its efforts on developing and disseminating disaster-resisting building best practices, training for labourers, and partnering with implementing organisations for the ret-rofit of damaged housing driven by the building owners themselves, referred to as owner-driven retrofitting (ODR) (Build Change, 2018).

In the context of disaster reconstruction, an effort to develop permanent housing is extremely challenging. The two different approaches employed by USAID and Build Change are very briefly described and compared below.

- Project team level diagnostic
  - Centrally planned interventions by large donors typically follow traditional implementation structures. The funder, acting as the owner, allocates a budget for the construction programme. It selects an implementing partner to act as its agent throughout the delivery process. The agent hires consultants to provide pre-construction, bid process and construction services, and hires a general contractor to execute the design. Upon completion and final approval, the agent turns the project over to either the funder or to a third party, such as a special-purpose entity or local government department, to manage the property and select beneficiaries.
  - ODR follows a significantly different trajectory. Funds are secured by an implementing organisation through grants or funder support. Beneficiaries are selected to receive cash grants, and the implementing organisation works with owners and local builders to provide training in retrofit best practices as well as engineering consultation and oversight during construction (Build Change, 2014).
- Service delivery diagnostic
  - The greenfield developments created new housing for beneficiaries directly, although work was planned in coordination with the Haitian public housing authority (the Construction and Housing Public Buildings Unit).
  - The services delivered in ODR are at the individual homeowner level, and therefore more closely resemble private sector approaches. Cash transfers to the beneficiaries mean that owners negotiate with providers directly. Build Change acts as the policy-maker in the process by establishing standards and providing engineering services to act as enforcement, while working directly with providers to make sure standards are met and sufficient technical capacity has been delivered.
- Process diagnostic
  - The implementation process of the centrally planned intervention more closely resembles traditional northern practice through planning, design and construction. US firms participate in many of the key roles in planning, design and construction. Local labour is employed during construction.
  - The project process in ODR is less formal. Full designs are not prepared for each house; instead, general guidelines are established and followed through construction (Guy Nordenson and Associates, 2011).
- Supply chain diagnostic
  - The centrally planned intervention is more likely to rely on non-local actors and materials throughout the process.
  - The owner-driven intervention relies more on local actors and materials. Build Change works closely with suppliers and labourers to make sure that specific standards are met and labourers possess adequate training in best practice (Build Change, 2012). Build Change engineers are employed to monitor progress. The primary risk to the success of the project in this case is quality, as it may be more difficult to ensure local builders are buying quality products if they are diffuse and rely on existing supply chains.

■ Regulatory and contextual diagnostic

– At the time of the earthquake, Haiti had no national building code. Buildings were built without the use of a building code, or, when buildings were designed by engineers, outside codes from the USA, Canada, the UK or France (and, later, Eurocode) were commonly employed. USAID required designs to meet International Building Code (IBC) standards, which is the US code.

– By 2013, Haiti had adopted a national building code that emphasises confined masonry design as the primary structural system and provides guidance on how to avoid vertical and horizontal irregularities. The guide is geared towards homeowners and master builders building without the assistance of professional architects and engineers, as this is the way the majority of buildings are constructed. The code references the 2009 IBC for all structures over three stories, those with irregularities or those employing a different structural system (e.g. reinforced concrete or steel) (Ministère des Travaux Publics, Transports et Communications, 2013). Build Change actively worked with the Government of Haiti in the development of the building code.

The discussion above illustrates two very different approaches to the same problem, providing safe housing for people affected by a natural disaster. The centrally planned intervention more closely resembles northern practice; however, ODR is more integrated with local practice. The success of either intervention relies heavily on the quality of the project teams and the programme design. Good civil engineering design is critical in both projects, with the centrally planned project favouring the IBC and the latter favouring confined masonry, an innovative and appropriate approach to earthquake-resistant retrofits. The scope of influence of the civil engineer in the centrally planned project, however, is limited to design and construction, so their ability to influence project success is limited. Civil engineering in ODR is embedded throughout the entire programme. Additionally, Build Change was able to influence national policy through demonstration of the pragmatic value of confined masonry construction and the development of a national building code that enshrines its practice. Both approaches contain their own challenges when bringing provision of shelter to scale. However, due to the lower cost of repairs over rebuilding and the reliance on local supply chains, ODR is potentially more easily scaled, particularly in dense urban environments (Build Change, 2014). While project processes that mirror northern project delivery practice are preferable to drive local adoption, centrally planned projects typically limit local participation in project development, such that very limited knowledge transfer is achieved, thereby eliminating any potential benefit of that approach.

The third case study, occurring several years after the first two, builds on lessons learnt from ODR. It demonstrates how donors and implementing organisations learn from the successful application of novel approaches. In 2016, Hurricane Matthew devastated the western coast of Haiti's southern peninsula. Catholic Relief Services, an implementing organisation tasked with immediate relief and reconstruction, integrated ODR into its relief plan, and, using other advancements in direct cash grants, developed a reconstruction programme that targeted improvements in the local supply chain and service delivery environment (CRS, 2019). The programme, called Salvage to Shelter, provided direct cash grants to households affected by the hurricane in the form of Q-code-enabled cards.

Beneficiaries were selected by local engineers carrying out post-event building surveys. Engineers prepared an assessment along with an estimate, and provided the homeowner with the money to rebuild in the form of credit. The organisation then worked with local suppliers to enable them to accept payment from homeowners and ensure supplies were available and of the necessary quality for reconstruction. The organisation also provided training for local builders in best practices. Homeowners were then able to purchase supplies and engage local builders in reconstructing their homes, with the organisation's engineers providing inspection and follow-up services. Vendors reported being able to more quickly establish their supply chains and reopen, while beneficiaries reported satisfaction in having control over their own reconstruction (Ward, 2018). This approach avoided the common issue of implementing organisations importing large quantities of foreign construction material, bypassing local suppliers and builders and creating shelters incompatible with local practice. Those projects typically end up providing immediate relief only and do not help affected areas build resiliency for future events.

Conclusions

The brief exploration of the examples above highlights three different approaches to the same humanitarian civil engineering intervention: shelter in post-disaster contexts. The role of the humanitarian civil engineer varies, from traditional design and construction services to working for an implementing organisation, supervising local engineers and training builders. The ODR model is an excellent example of how humanitarian civil engineers can help develop new approaches to interventions that emphasise local participation, build capacity and strengthen local supply chains. The ODR and Salvage to Shelter programmes both address root causes of problems.

The diagnostic framework provided above can be a useful tool to help humanitarian civil engineers diagnose and analyse a project or programme to (a) provide recommendations for executing a strategy that emphasises northern delivery practices and local participation, and builds capacity for various entities and groups throughout the service delivery process and supply chain; (b) redefine the role of the humanitarian civil engineer within that context, for projects with clearly defined boundaries; and (c) reimagine the scope or programme to address the root causes of lack of infrastructure provision and services in the first place.

Humanitarian civil engineering practitioners will be engaging in a dynamic environment that continues to change as the challenges facing societies evolve and attitudes towards international cooperation and development shift. For young and old, practice is rooted in professional mastery of a core discipline and the attendant tasks and skills. Hopefully, this sub-chapter provides some grounding to help understand the humanitarian milieu and place oneself within it, as well as provide ideas for further study, and arms the reader with guidance to successfully diagnose a project or project environment to determine the appropriate role for the humanitarian civil engineer within the project context. Humanitarian civil engineers are their own greatest advocate for the importance of civil engineering in project delivery, and should avoid taking roles that undermine the development of robust infrastructure service delivery within the contexts where they work.

REFERENCES

ALNAP (Active Learning Network for Accountability and Performance) (2009) *Participation Handbook for Humanitarian Field Workers*. ALPNAP, London, UK.

Bailey S and Harvey P (2017) *Time for Change: Harnessing the Potential of Humanitarian Cash Transfers*. Overseas Development Institute, London, UK.

Bradshaw S and Fordham M (2013) *Women, Girls and Disasters: A Review for DFID*. Department for International Development, London, UK. https://assets.publishing.service.gov.uk/government/uploads/system/uploads/attachment_data/file/844489/withdrawn-women-girls-disasters.pdf (accessed 12/04/2021).

Buchanan-Smith M and Maxwell S (1994) Linking relief and development: an introduction and overview. *IDS Bulletin* **25**: 2–16, 10.1111/j.1759-5436.1994.mp25004002.x.

Build Change (2012) *Challenges and Opportunities in the Production and Purchase of Good Quality Blocks*. Build Change, Denver, CO, USA.

Build Change (2014) *Homeowner-Driven Housing Reconstruction and Retrofitting in Haiti*. Build Change, Denver, CO, USA.

Build Change (2018) *Success in Haiti: Achievements from 2010 to 2018 (and Beyond)*. Build Change, Denver, CO, USA.

Cabannes Y (2014) *Contribution of Participatory Budgeting to provision and management of basic services: Municipal Practices and Evidence from the Field*. International Institute for Environment and Development, London, UK.

CARE India (2016) *Post-disaster Shelter in INDIA: A Study of the Long-term Outcomes of Post-disaster Shelter Projects*. CARE India, New Delhi, India. https://insights.careinternational.org.uk/media/k2/attachments/Post-Disaster-Shelter-in-India_full-report_2016.pdf (accessed 12/04/2021).

Chmutina K, Von Meding J and Bosher LS (2019) Language matters: dangers of the 'natural disaster' misnomer. In *Global Assessment Report on Disaster Risk Reduction 2019*. United Nations Office for Disaster Risk Reduction, Geneva, Switzerland. https://www.preventionweb.net/publications/view/65974 (accessed 12/04/2021).

Cooper AJ (1892) *A Voice from the South*. Aldine, Xenia, OH, USA.

Cordero-Scales C, Kellum J, Rule A and Harriss L (2016) *Gender and Shelter: Good Programming Guidelines*. CARE International, London, UK.

Crawford K (2018) *Danger! Weird ways engineers think and talk about disasters in cities*. Resilient Urbanism. https://resilienturbanism.org/kcrawford/danger-weird-ways-engineers-think-and-talk-about-disasters-in-cities/ (accessed 12/04/2021).

Crenshaw K (1989) Demarginalizing the intersection of race and sex: a Black feminist critique of antidiscrimination doctrine, feminist theory and antiracist politics. *University of Chicago Legal Forum* **1989(1)**: 139–167.

Crenshaw K (1991) Mapping the margins: intersectionality, identity politics, and violence against women of color. *Stanford Law Review* **43(6)**: 1241–1299, 10.2307/1229039.

Criado Perez C (2019) *Invisible Women: Exposing Data Bias in a World Designed for Men*. Chatto and Windus, London, UK.

CRS (Catholic Relief Services) 2019. *Haiti: Cash-Based Interventions in Shelter: The Vendor Effect*. CRS, Baltimore, MD, USA.

Davis J and Lambert B (2002) *Engineering in Emergencies*, 2nd edn. Practical Action Publishing, Rugby, UK.

GFDRR (2017) *Building Back Better in Post-disaster Recovery*. GFDRR, Washington, DC, USA. https://www.recoveryplatform.org/assets/tools_guidelines/GFDRR/Disaster%20Recovery%20Guidance%20Series-%20Building%20Back%20Better%20in%20Post-Disaster%20Recovery.pdf (accessed 12/04/2021).

Guy Nordenson and Associates (2011) *Design and Construction Guidelines for Confined Masonry Housing*. Guy Nordenson and Associates, New York, NY, USA.

Guy-Sheftall B (1995) *Words of Fire: An Anthology of African-American Feminist Thought*. The New Press, New York, NY, USA.

ICE (Institution of Civil Engineers) (2004) *ICE Code of Professional Conduct*. ICE, London, UK. https://www.ice.org.uk/ICEDevelopmentWebPortal/media/Documents/About%20Us/ice-code-of-professional-conduct.pdf (accessed 12/04/2021).

IFRC (International Federation of Red Cross and Red Crescent Societies) (2018a) *Case Study: Integrating CLTS and PHAST in Kenya*. IFRC, Geneva, Switzerland. https://media.ifrc.org/wp-content/uploads/sites/13/2018/10/ifrc_clts-phast-case-study_kenya_en.pdf (accessed 12/04/2021).

IFRC (2018b) *IFRC Policy Brief. Localization: What it Means and How to Achieve It*. IFRC, Geneva, Switzerland. https://media.ifrc.org/ifrc/document/ifrc-policy-brief-localization/ (accessed 12/04/2021).

Institution of Structural Engineers (2020) *Working in the humanitarian or development sectors*. https://www.istructe.org/resources/guidance/working-in-humanitarian-development-sector/ (accessed 12/04/2021).

Kennedy J and Newby T (2018) Just one small part of the jigsaw: why shelter response must serve complicated human REALITIES. In *The State of Humanitarian Shelter and Settlements 2018. Beyond the Better Shed: Prioritizing People*. Global Shelter Cluster, ch. 8. https://www.sheltercluster.org/resources/library/state-humanitarian-shelter-and-settlements (accessed 12/04/2021).

Klein N (2007) *The Shock Doctrine*. Allen Lane, London, UK.

Krishna A (2003) Partnerships between local governments and community based organizations: exploring the scope for synergy. *Public Administration and Development* **23**: 361–371, 10.1002/pad.280.

McCall L (2005) The complexity of intersectionality. *Signs* **30(3)**: 1771–1800, 10.1086/426800.

Macdonald I (2020) *10 years ago, we pledged to help Haiti rebuild. Then what happened? In These Times*. https://inthesetimes.com/features/haiti_earthquake_recovery_us_aid_anniversay_military_waste.html (accessed 12/04/2021).

Marcussen HS (1996) NGOs, the state and civil society. *Review of African Political Economy* **23(69)**: 405–423, 10.1080/03056249608704205.

Ministère des Travaux Publics, Transports et Communications (2013) *Code National du Batiment d'Haiti 2012*. Bibliotheque Nationale d'Haiti, Port au Prince, Haiti.

Mohan G and Stokke K (2000) Participatory development and empowerment: the dangers of localism. *Third World Quarterly* **21(2)**: 247–268, 10.1080/01436590050004346.

Mughal H, Ahmed SA, Mumtaz H, Tanwir B, Bilal S and Stephenson M (2015) *Kashmir Earthquake 2005: Learning from the Shelter Response and Rural Housing Recovery*. International Organization for Migration and Shelter Cluster, Geneva, Switzerland. https://www.sheltercluster.org/sites/default/files/docs/shelter_in_recovery._kashmir_eq_2005.pdf (accessed 12/04/2021).

ODI (Overseas Development Institute) (2015) *Doing Cash Differently: How Cash Transfers Can Transform Humanitarian Aid. Report of the High Level Panel on Humanitarian Cash Transfers*. ODI, London, UK. https://odi.org/en/publications/doing-cash-differently-how-cash-transfers-can-transform-humanitarian-aid/ (accessed 12/04/2021).

Sanderson D and Sharma A (2008) Winners and losers from the 2001 Gujarat earthquake. *Environment and Urbanization* **20(1)**: 177–186, 10.1177/0956247808089155.

Schumacher EF (1973) *Small is Beautiful: Economics as if People Mattered*. Harper and Row, New York, NY, USA.

Slim H (2015) *Humanitarian Ethics: A Guide to the Morality of Aid in War and Disaster*. Oxford University Press, New York, NY, USA.

Smith S (2013) Black feminism and intersectionality. *International Socialist Review* **91**: 6–24. https://isreview.org/issue/91/black-feminism-and-intersectionality (accessed 12/04/2021).

Twigg J (2015) *Disaster Risk Reduction: Mitigation and Preparedness in Development and Emergency Programming. Good Practice Review 9*. Humanitarian Practice Network, London, UK.

UN (United Nations) (2015) *Sendai Framework for Disaster Risk Reduction 2015–2030*. UN, Geneva, Switzerland. https://www.preventionweb.net/files/43291_sendaiframeworkfordrren.pdf (accessed 12/04/2021).

UNISDR (2017) *Build Back Better: In Recovery, Rehabilitation and Reconstruction. Consultative Version*. UNISDR, Geneva, Switzerland. https://www.unisdr.org/files/53213_bbb.pdf (accessed 12/04/2021).

US Office of Inspector General (2014) *Audit of USAID/Haiti's New Settlement Construction Activities*. US Agency for International Development, San Salvador, El Salvador.

USAID (US Agency for International Development) (2015) *Fast Facts on the U.S. Government's Work in Haiti: Shelter, Housing, and Settlements*. USAID, Washington DC, USA. https://2009-2017.state.gov/documents/organization/248752.pdf (accessed 12/04/2021).

Ward S (2018) *The Vendor Effect: Hurricane Matthew Response. A Cash Transfer Programing Learning Study*. CRS Haiti Catholic Relief Services, Baltimore, MD, USA.

Whaites A (1998) NGOs, civil society and the state: avoiding theoretical extremes in real world issues. *Development in Practice* **3(8)**: 343–349, 10.1080/09614529853639.

White H, Menon R and Waddington H (2018) *Community-Driven Development: Does it Build Social Cohesion or Infrastructure? A Mixed-Method Evidence Synthesis. International Initiative for Impact Evaluation. Working Paper 30*. International Initiative for Impact Evaluation, New Delhi, India.

World Bank (1990) *Strengthening Local Governments in Sub-Saharan Africa*. The World Bank, Washington, DC, USA.

World Humanitarian Summit (2016) *The Grand Bargain: A Shared Commitment to Better Serve People in Need*. Agenda for Humanity, Istanbul, Turkey.

Zarins J, Newby T and Haiselden S (2018) What does this mean for the shelter sector? In *Lessons from Typhoon Haiyan: Briefing. A Review of Shelter Self-recovery Projects in the Philippines, and their Lessons for the Shelter sector*. CARE International, London, UK, ch. 2. https://insights.careinternational.org.uk/publications/lessons-from-typhoon-haiyan-briefing (accessed 12/04/2021).

Georgia Kremmyda
ISBN 978-0-7277-6468-3
https://doi.org/10.1680/hce.64683.025
ICE Publishing: All rights reserved

Chapter 2
Civil engineers and humanitarian challenges

Abstract

The chapter aims at building awareness of the extent and scale of the world's humanitarian needs, and the call to action for civil engineers in addressing humanitarian challenges and ensuring sustainable development. The chapter starts with illustrating the application of civil engineering in humanitarian relief settings. Examples are drawn from construction projects in the Democratic Republic of Congo (DRC), where the general principles of applying civil engineering in humanitarian settings are highlighted in comparison with conventional engineering. Within the humanitarian engineering context, the chapter examines some of the challenges that built environment professionals face, predominantly civil engineers and architects. A discussion, based on extensive experience of the CARE International Emergency Shelter Team, unfolds on the role of civil engineers in supporting the re-establishment of housing and communities after catastrophic events such as storms, earthquakes and floods, as well as the needs of communities affected and often displaced by conflict. Like conventional civil engineering projects, those for humanitarian work must equally balance economy, programme and quality. The importance of maintaining quality for the long-term benefit of the beneficiaries is paramount.

APPLYING CIVIL ENGINEERING IN HUMANITARIAN SETTINGS

Simon Bird and Gustavo Cortés

Introduction

The value of engineering in a humanitarian setting is well illustrated in the origins of the disaster relief charity RedR. Engineer Peter Guthrie worked in a refugee camp, where there was a high incidence of waterborne diseases. The problem was addressed by improvements in the water system, while other professionals were seeking curative solutions. Peter then had the idea for RedR, to facilitate engineering assistance in such settings (RedR, 2021).

On the other hand, engineers should not overrate their value. A typical humanitarian project draws together many disciplines, such as management, finance, logistics, human resources and monitoring. Humanitarian interventions are often simple and repeated, for example the distribution of tarpaulins as emergency shelter provision, food distribution or the provision of tap stands; they do not require significant technical input. However, some details may be crucial, such as the specification of the tarpaulins, including their density and the number of eyelets, and the instructions on how to fix them in order to prevent premature failure. Any construction involved is usually relatively simple (e.g. for well protection, latrine covers and clinic buildings) – within the capacity of unskilled, but supervised, local labour. Nevertheless, when technical decisions are made it is important that they are correct, otherwise they will lead to major consequences in the long term. For example, Medair's reconstruction work in rural Nepal after the 2015 Gorkha earthquake required that houses be rebuilt using a higher earthquake engineering standard, to avoid rebuilding the same vulnerabilities that caused great devastation in the first place.

Engineering education focuses on teaching students how to learn – how to be problem-solvers. Although very specific problems and technologies are presented throughout their coursework (e.g. the design of a wood beam), the nature of the problems should not be the main focus, but rather the methodologies followed to solve it. In that sense, engineers normally have fully transferable skills that can easily be applied in humanitarian contexts.

However, being a project manager in the humanitarian sector can be very challenging for young engineers trained to be problem-solvers. The challenge arises from that same desire to solve problems, which are often not very clearly defined. In the humanitarian sector much community work needs to happen, since the community knows best its needs, and may know many effective solutions that accord with local conditions. In other words, as problem-solvers, engineers might be tempted to find a solution to a problem that may not be the real problem the community is facing, or which may not be culturally appropriate. Thus, engineers need to listen first, and involve the community to determine the needs and the solutions: a skill that is learned in the field.

New people in the field may be surprised about similarities between humanitarian and engineering in the industrial world. It is still necessary to balance budget, programme and quality. Obtaining funding often involves proposals with a well-thought-out rationale, illustrated by drawings, specifications and bills of quantity. The work and necessary labour and material are planned using typical project tools. The nature of the work is like that of a small building contractor. So, engineers used to engineering in the industrial world do have many useful and varied transferable skills.

The application of civil engineering in humanitarian settings will be illustrated in this sub-chapter chiefly by reference to Medair's work in the DRC.

Medair's work in the DRC

General

Medair is a humanitarian organisation inspired by Christian faith to relieve human suffering in some of the world's most remote and devastated places (Medair, 2021). Medair has worked in the DRC for some 30 years, addressing health, nutrition, water, sanitation and hygiene (WaSH), and infrastructure needs. At the time of writing, it was also responding to the Ebola crisis.

Ground communication in the DRC has always been difficult. The terrain and climate present challenges to construction, and distances are large. Road and bridge infrastructure built in the colonial era is in very poor condition, suffering years of neglect due to conflict, political instability and poor governance (World Bank, 2021). In the areas in which Medair works, virtually no maintenance is carried out by the responsible body, so villagers repair roads and replace bridge decks with locally cut tree trunks (Figures 2.1 and 2.2).

Figure 2.1 Medair's vehicle having difficulties through rough unpaved roads in the DRC

Figure 2.2 Lorry accident after a locally made tree trunk bridge failure

In a health project in North Kivu Province it was apparent that even pedestrian access to present and planned clinics was very poor. Local bridges were constructed of timber and supported by creepers (Figure 2.3). These sufficed for fit persons, but were unsuitable for carrying in building and medical supplies or for people on stretchers. During the project-planning phase, two people died: one of malaria because the old clinic was not stocked with the relevant drugs, and the other, a woman in childbirth, because she could not be carried to the clinic in time. The bridges are also susceptible to flooding, so more robust and higher suspended bridges were planned.

Medair has worked in infrastructure throughout the DRC, on the construction of reinforced concrete bridges, the refurbishment of Bailey bridges and building a number of pedestrian suspension bridges. Two recently completed bridges are discussed in detail below: the first is a suspended pedestrian bridge, the second is a two-span road bridge. A very brief description of the technical design is provided, and the challenges experienced are also described.

Suspended bridges

Medair has built several suspended bridges following the Bridges to Prosperity guidelines (B2P, 2014). These bridges are simple in principle and have the advantage of not using piers in the rivers. There are four main components to these bridges: the cables, the towers, the anchors and the deck. The design is relatively simple, determined by basic statics, the construction is relatively straightforward and the cost is attractive. Figure 2.4 shows a free-body diagram of the tower and the anchor. From this diagram it is relatively simple to determine the demand on the main cables supporting the live and permanent loads, on the tower and its foundations, and on the anchor, which is designed for uplift and sliding. Once these demands are established, all members can be designed. Full explanations on this process can be found in the B2P guidelines. Figure 2.5 shows a typical suspended bridge built by Medair.

Figure 2.3 Locally built pedestrian bridge

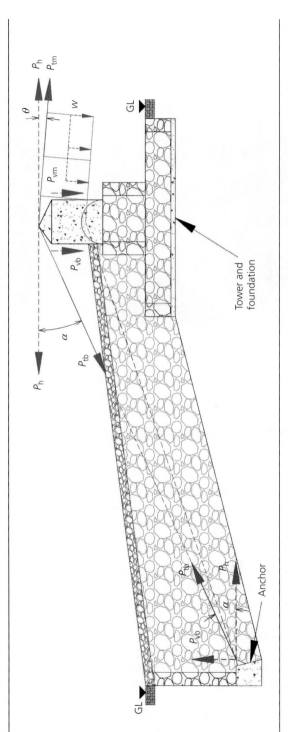

Figure 2.4 Free-body diagram of the tower and the anchor

Figure 2.5 Suspended bridge constructed

Procurement was eventful. Initially the logistics team in the DRC was unfamiliar with the nature of the cables needed. It sent brochures on cables to its headquarters, but these had very low tensile strength and turned out to be aluminium conductor cables. It then set about obtaining steel cables from Uganda. However, it transpired that steel cables were available for purchase in the DRC, and it is believed that donations could have been obtained from the mining companies in the project locality. B2P advises that it obtains all its cables as donations from the shipping industry.

Similarly, the recent bridge in Hombo required five 24 mm diameter cables. However, after much searching, only 26 mm diameter cables could be found. Thus, the team decided to go with these since they had higher capacity. However, once the cable was received, the chosen cable was found to be very rigid, making the required bending around the anchors difficult. This shows that diameter and breaking strength are not the only two factors to consider when choosing the right cable.

Transporting the cables to the site proved challenging. Cables were delivered by lorry to Goma, then by car or helicopter to the nearest major town, and finally by a team walking for sometimes as much as a week to bring the cables to the construction site (Figure 2.6).

Two-span road bridge in Taliha

Medair's first road bridges were built in response to a need seen by several non-governmental organisations (NGOs) working in Dingila, Orientale Province, DRC, to improve access for many humanitarian projects, described by Bird and Simon (2013). No contractors existed in this region, so, as Medair had experience of construction of latrines and clinics, they took on this work, which was funded by the United Nations Development Programme pool fund. Further funding was obtained for more bridges, to improve access generally for beneficiaries in the region. Figure 2.7 shows one of the first bridges that Medair constructed.

Figure 2.6 Supply of cables to a remote area inaccessible to vehicles

Figure 2.7 One of the first road bridges completed by Medair in the DRC

In 2019, Medair built a bridge in Taliha, to access an area with a significant cholera outbreak, where crossing the river to bring medicines was difficult and risky.

The local community had started to build a bridge with the intention to use tree log beams, but was unable to finish it due to budgetary constraints. Given that this was the main access road to Medair's projects, and essential for the people in this community to access health services, Medair decided to finish the bridge. The two-span steel beam bridge consisted of a reinforced concrete central pier and two cyclopean concrete (mass-concrete-containing boulders) abutments. The abutments had been built, but Medair had to reinforce one and completely rebuild the other due to poor quality. The initial design (Figure 2.8) was adjusted to suit the steel sections found in Goma, DRC. The completed bridge is shown in Figure 2.9.

Transporting the steel sections, and other construction materials such as cement, was difficult. The project site could be accessed only through very narrow off-road motorcycle paths, making it very dangerous and time consuming to bring materials.

One significant challenge was that, due to an error in supervision, the pier was not built in the centre of the bridge. Changes to the beams had to be made on site. The solution was to splice beams at specific locations and to connect the beams using welded plate splice connections, as shown in Figure 2.10. In addition, the two centre beams were welded along the flanges using intermittent fillet welds, to ensure that the full capacity of the section was achieved at the splices.

General principles

General

Having described specific bridge projects in the DRC, more general principles of applying civil engineering in humanitarian settings are drawn out in this section, to illustrate the nature of civil engineering in humanitarian settings.

Constraints

It is difficult to generalise, but the context of humanitarian work is likely to be more challenging than that of engineering in developed countries. The main cause is the urgency to deliver, given that the community being served is often in dire conditions, usually a result of a natural disaster or armed conflict. The challenges apply to the conditions in which engineers must work, the feasibility of executing the work, and the nature of solutions adopted. The constraints within which humanitarian engineering is practiced are typically shortage of funds, difficulties of physical access due to shortcomings of infrastructure, and difficulties of access due to security, political and cultural constraints. There may be pressure for rapid emergency interventions. Also, trained design and construction staff, construction equipment and quality-assured materials may not be available.

Appropriate technology

Humanitarian responses do not necessarily call for high-technology solutions. The technical content is generally quite basic, but nevertheless is important to get right. Good engineering is always about producing the best solution within the project constraints that benefits the end

Figure 2.8 Taliha bridge side elevation

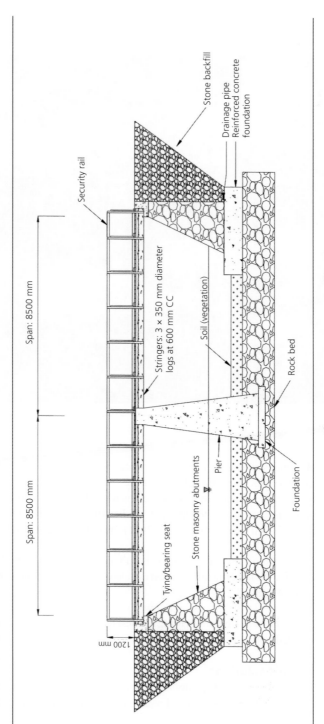

Figure 2.9 Taliha bridge

users, and makes the best uses of available resources, whether it is a major dam or a spring protection. It is axiomatic from the Charter of the Institution of Civil Engineers that civil engineering is 'for the use and convenience of man' and takes a central role in building sustainable economies and stable governments (ICE, 2016). Also, the code of ethics of the American society of Civil Engineers (ASCE, 2021) says in its first canon that

> Engineers shall hold paramount the safety, health and welfare of the public and shall strive to comply with the principles of sustainable development in the performance of their professional duties.

Thus, good engineering is always about using the best technology appropriate to the setting, as will be demonstrated below.

Simple and robust solutions

Lack of a skilled workforce and quality-assured materials, as well as the need to provide solutions that can be easily maintained and repaired by locals, mean that solutions should be

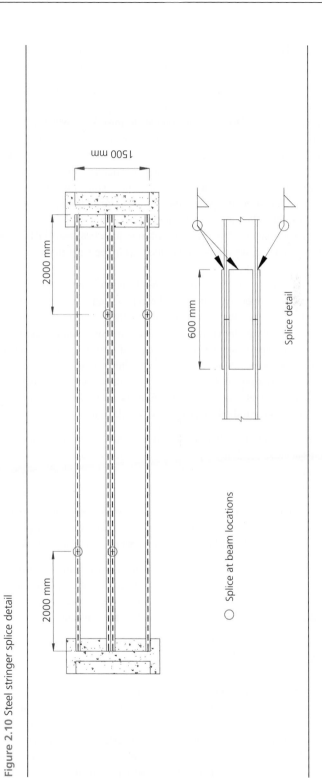

Figure 2.10 Steel stringer splice detail

1500 mm

2000 mm

2000 mm

600 mm

Splice detail

○ Splice at beam locations

simple and robust. The settings in the DRC are very remote, and no contractors or other local expertise exists. Examples of simple solutions include the following.

- Bridge decks were designed as simply supported, to avoid complex reinforcement for the concrete bridges and moment transfer to masonry abutments.
- Concrete decks were flat slabs, to avoid complex formwork. As experience was gained, more adventurous solutions were adopted: composite decks and precast beams with deck infill.
- Bracing to reduce the unbraced length of the steel beams stringers was avoided as being difficult to fabricate, with two I-shaped beams welded together intermittently used at the centre instead. Welding the two beams together avoided lateral–torsional buckling and allowed the full strength of the cross-section. The additional capacity also increased the margin of safety, allowing for unanticipated loading.
- Bridges were short enough (<6 m) so as not to require bearings (TRL, 2000). As experience was gained, movement joints made with recycled rubber tyre inner tubes were used in longer bridges.
- Abutments were mass concrete. This avoided placing reinforcement in foundations, which were prone to flooding, and avoided complex formwork and reinforcement fixing.
- Mass concrete also suited the refurbishment of existing abutments, which were dry-stone masonry. Skills for cutting and placing masonry did not exist, so existing foundations were reused, and the stones recovered from the collapsed stonework was reused as core stones in the mass concrete, thus saving cement and aggregates.
- Standardisation of design. As far as possible, the same cross-sections for the deck, abutments and wing walls, details and reinforcement layouts were used, to ease supervision time and avoid mistakes by unskilled labour.

It is also important to note that, like most NGOs, Medair trains local labour to increase its skills. This helps achieve the desired quality in the projects and also helps people to get better jobs later on. In the DRC, employing local labour for each bridge had many social benefits: it gave skills in construction, increased self-respect and provided additional local income.

In industrialised countries, with high labour costs and efforts to design-out operations in hazardous locations, solutions are characterised by prefabricated or labour-saving components, which may increase material costs, but reduce labour and the number of operations by various trades.

In contrast, the use of imported materials (e.g. cement and steel reinforcement) was minimised in the DRC, to reduce the relatively high cost and the difficulty of transport to the remote locations. The cost of local labour is very low, and there were social advantages to maximising the use of local labour, which was very willing but largely unskilled. Examples include

- Cutting local timber for use in formwork, suspended bridge and refurbished Bailey bridge decks.
- Using locally quarried stone and aggregate.
- Minimising reinforcement in the design and detailing – especially as, in some locations, reinforcement had to be delivered by light aircraft – by, for example, detailing lengths to avoid waste bars, and placing the exact number of bars needed, rather than rounding spacing to the nearest 25 mm. In the circumstances it was decided to omit top reinforcement for simply supported decks with sufficient bottom steel for strength and service considerations – although this has benefits it is not considered essential (IStructE, 2006).

A number of types of bridge were considered for the footbridges (e.g. timber truss and steel girder). However, mainly because of the span, and height needed to clear flood water, suspended bridges were chosen. Although steel cables were required to be imported, this was a relatively small item, and the other components could be largely sourced locally.

Design process

All civil engineering design involves collaboration between parties and consultation with stakeholders. Humanitarian settings are no exception. Perhaps the greatest cultural shift that an engineer from outside must make is to learn to make the best use of local knowledge and resources and to communicate with stakeholders who may be scattered across the globe. Some stakeholders in the design process are

- Beneficiaries. As well as being a requirement of Sphere standards (Sphere, 2018) and Core Humanitarian Standards (CHS Alliance, 2014), consulting and involving beneficiaries in the design process will greatly improve success. They are effectively the end users, and will have valuable suggestions on what works well, the use of local materials, and techniques that are suited to the conditions. As well as the social benefits of employing local labour, this also helps to secure acceptance of the project, participation in the project formulation and later maintenance. It also assists in ensuring the security of staff and the works, which might not be the case with labour from outside.
- Local engineers. In many countries of operation there are often technically trained staff, highly motivated to help their people, among the affected populations, possibly with degree-level training and construction experience. However, in some places where Medair works this has not been the case. There are also instances when there are staff whose construction experience is mainly from work in cities or on major schemes, so they too must learn to adapt to humanitarian settings as much as expatriate engineers.
- Local supervisors. Competent local construction supervisors are probably the most important key to success. In the DRC, Medair had no experience of major construction – but it did have experience of the construction of clinics, wells and latrines, so the supervisors had basic construction skills to build on. They were guided by the designer in improving key skills, as discussed below.
- The designer. With global communication, the designer does not need to be based in the same country as the project, perhaps in a remote national office. Design for the bridges in the DRC was carried out remotely, so there had to be frequent and close communication between the local staff and the designer, with efficient transfer of drawings and other information. It is encouraged, however, that the designer visits frequently.
- Headquarters. Checking designs, providing an extra layer of quality assurance, setting policy, making standards and enforcing them, training staff, and verifying compliance with donor requirements and minimum standards are done by the headquarters.
- Donors. Review of designs at the proposal and other stages is done by the donors. They visit the projects when security allows.

As with local labour, Medair seeks to train and expand the horizons of locally recruited engineers and other technical staff. Medair holds annual sector conferences, including one for shelter and infrastructure, where locally recruited national staff gather at a regional centre to present their work and receive training.

Another source of design is commercial companies. Many firms will carry out work for no or reduced fees under corporate social responsibility (CSR) schemes. They provide a good introduction for younger engineers to learn about the humanitarian sector, and enable humanitarian organisations to tap into the wide experience and design resources of a major design company. The chief issue noted by the authors is lack of awareness in the field of potential assistance from outside, issues of setting commercial arrangements, the definition of problems and developing suitable work packages.

It is important to note that safety in construction is a prime consideration, but a poor safety culture may exist. Medair will avoid dangerous practices and seek to improve awareness of safety through discussion with communities.

Use of standards

In the construction of the bridges in the DRC, Medair followed standard designs as far as possible: *Overseas Road Note 9* (ORN9) (TRL, 2000) for road bridges, and *Bridges to Prosperity Suspended Bridge Manual* (B2P, 2014) for suspended bridges, but in both cases it had to adapt the standards according to the circumstances. This required engineering skill and judgement, and illustrates the kind of decisions that may need to be taken.

It was not possible to follow ORN9 in a number of ways (fully described by Bird and Simon, 2013), as described below.

- Because of the difficulty of access, and security constraints, the bridge sites could be visited only briefly over one day, so it was impossible to gather information on the river profile upstream and downstream. In mitigation, the land was generally flat, so river flows had limited potential for scour, and bridges were largely replacing existing bridges, so it was assumed that the existing layouts were satisfactory (Figure 2.11).
- Again because of access constraints, and because it would have been prohibitively expensive, an intrusive ground investigation was not possible. In order to avoid an excessively conservative design, allowance was made for assessing foundations at the time of construction, and the foundations could be widened, and shear key added, according to conditions. This was another advantage of adopting simple mass concrete abutments and wing walls – they could be adjusted more easily than reinforced concrete walls (Figure 2.12).
- Despite extensive enquiries, guidance from the local road authorities on codes of practice, traffic and loading was not obtained until a late stage in the project. Lorries that use the road were investigated and the loading was calculated, including an allowance for overloading. This was less than UK Highways Agency loading (UK Highways Agency, 2001). Incidentally, on this basis the designs of ORN9 were approximately 30% overdesigned. Thus, a degree of conservativism was introduced that mitigated for many of the unknowns, but not excessively so.
- In the absence of local guidance, structural design was carried out to British Standards, Eurocodes or other internationally accepted codes. For example, Medair followed the International Building Code (ICC, 2018) for the shelter response following the 2010 earthquake in Haiti, and a Nepal Building Code (Government of Nepal, 2015) following the Nepal earthquake.

Figure 2.11 Survey at the bridge location

Figure 2.12 Mass concrete abutments

The *Bridges to Prosperity Suspended Bridges Manual* (B2P, 2014) was developed in Nepal, and it was very useful to have its extensive design guidance with detailed, practical and well-thought-out standard designs. But these were adapted in a number of respects.

- Because the cables were imported, it was desired to use the smallest number and size consistent with safety and durability. The B2P loading recommendations were carefully examined (Table 2.1). The B2P loading is equivalent to four people per metre along the entire length of a bridge. Since the DRC bridges would be only 1 m wide, this represents an extreme crowd loading. The EC1 loading is equivalent to one person per metre. The DRC bridges would be in remote areas, and the population is already used to one person at a time on the present very flimsy timber bridges, so the EC1 loading was chosen as being appropriate, and warning signs were added to avoid more than one person, one motorbike or one person being carried on a stretcher at a time. However, at the opening ceremony of Taliha bridge, discussed above, the bridge was packed, and the additional capacity of the doubled beams was a comfort. Thus, engineering judgement is paramount, and deviation from standards is only advised when there is a strong conviction founded on facts.
- The designs of the B2P bridges showed concrete or masonry towers. In order to save on use of cement, for some of the DRC bridges the towers were made of gabions, which maximised the use of local materials (Figure 2.13).

Quality

Like conventional civil engineering projects, those for humanitarian work must equally balance economy, programme and quality. Because of difficulties of access and supply and the urgency of a situation, there is often pressure to 'cut corners' or adopt substandard solutions. The same is true in disaster relief, but the balance will vary because of constraints. Quality is often neglected, but, as is impressed on Medair staff

> When we have all gone home, no one will care if the bridge was slightly over cost or late (of course we do try to avoid these); what the beneficiaries will notice is the quality – is the work durable, and still meeting their needs?

Table 2.1 Comparison of loading from various standards

Standard	Recommended distributed load: kN/m^2
Bridges to Prosperity Suspended Bridges Manual: live loads, suspended bridges	3.00
British Standard BS CP3: footpaths, terraces and plazas	4.00
Eurocode 1 (EC1, BS EN 1991): areas where people may congregate – public access areas	3.00
EC1: construction loads, working personnel with hand tools	0.75

Figure 2.13 Suspended bridge gabion tower

We have found mistakes in quality that can have a major impact. In one country, difficulties in supervision allowed poor-quality latrine soakaways to be constructed. Returning to repair them was expensive.

In the DRC the supervisors were experienced in building clinics, wells and latrines. However, they were not used to providing concrete of load-bearing standard. There were a number of areas where the designer had to work with them to provide assurance of complying with the structural standards.

Quality-assured materials cannot normally be obtained, in contrast to an industrialised setting, where concrete, for example, can be specified and delivered from a batching plant, with certificates of assurance. Concrete in the field can often only be made by hand (Figure 2.14(a)) using local sand and aggregates, and testing is not feasible. So, it is necessary to make judgements about material parameters for use in design, and set ways of controlling the materials, as the following examples from the DRC show.

- A cautious concrete cube strength for design of 25 MPa was used.
- The designer gave succinct instructions on what was necessary to obtain good concrete (e.g. obtaining clean aggregates, and careful volume measurement, mixing, placing and hand compacting). A particular problem was avoiding segregation by moving concrete around in the formwork and adding too much water.
- Steel reinforcement was assumed to be mild steel.
- Attention was given to the correct positioning of the formwork and reinforcement.
  At first, stones were used to provide correct cover, but later the workforce was taught to make cover blocks (Figure 2.14(b)).

- Local timber was used for temporary diversion road bridges, for replacement Bailey bridge decks and for the deck of the suspended bridges. It was therefore necessary to ascertain the timber grade for use in design – but only the local name of the trees used was known. Therefore, a simple test for a timber beam was improvised – loading with buckets of aggregate and measuring deflection – to calculate Young's modulus, and hence to determine the grade for use in design codes.
- Another challenge was the quality assurance programme, as access to the necessary information by Medair staff was frequently not possible due to insecurity. Thus, quality is often in the hands of a locally recruited foreman, who, speaking the local language and trusted, may safely stay in the area. For this reason, it is essential that the foremen are well trained and are committed to ensuring the required quality.

Figure 2.14 (a) Hand mixing concrete. (b) Reinforcement cover block

(a)

(b)

Conclusions

Two bridges constructed in challenging conditions in the DRC have been described. They illustrate some of the constraints of civil engineering work in humanitarian settings. Engineers familiar with standards, quality-assured materials and ease of procurement must make cultural shifts and adjustments to the prevailing conditions. They must use technology appropriate to the setting, adopt simple and robust solutions, make best use of local knowledge and resources, and communicate with stakeholders who may be scattered across the globe. They should use conventional standards of design, but be prepared to intelligently adapt these to the conditions. Like conventional civil engineering projects, those for humanitarian work must equally balance economy, programme and quality, and the importance of maintaining quality for the long-term benefit of the beneficiaries has been emphasised.

The bridge projects described provide humanitarian access to people in remote locations, where the lack of road and bridge infrastructure often results in difficulties when bringing medicine to health and nutrition facilities, and providing access so that the sick may reach such facilities. These bridges also interconnect communities, providing access to children to go to school and to families to access markets that otherwise would not be possible.

The authors have been privileged to apply their engineering knowledge gained in industrial settings for the use and convenience of beneficiaries facing difficult circumstances. They encourage other engineers to enter the humanitarian field.

Acknowledgements

The work described in this sub-chapter was carried out by Medair DR Congo with funding from various sources. The authors are grateful to Medair for permission to publish this work. The views represented here are those of the authors and not necessarily those of Medair.

IS SAFE THE ENEMY OF SAFER? HOUSING AFTER DISASTERS AND THE ROLE OF BUILT ENVIRONMENT PROFESSIONALS?

Bill Flinn and Step Haiselden

Introduction

Tropical Cyclone Pam swept through the Pacific island nation of Vanuatu on 13 March 2015. Sustained windspeeds were estimated at 250 km/h with gusts peaking at 320 km/h, making it the most powerful storm ever recorded in the region. Vanuatu is considered one of the most environmentally exposed nations in the world, prone as it is to volcanic eruptions, cyclones, earthquakes and the effects of climate change.

The southern island of Tanna took the brunt of the cyclone. The destruction in the villages was devastating. In many, up to 80–90% of the pole-and-thatch houses were completely destroyed. It is a reflection of the extraordinary resilience and preparedness of the ni-Vanuatu people that very few died – 11 is the official figure (OCHA, 2015).

Even amid the debris of destruction, two factors were immediately apparent. Some traditionally built houses were still standing and, knowing that they were strong, the villagers had used these houses to take refuge in during the night of the storm. Local knowledge saved lives. Second, by the time the first aid workers were able to visit the villages, the remains of houses were already being salvaged; the thatch from the roofs was being laid out to dry and rebuilding was under way. Here was proof that people are never passive and are always the first and most important actors in their own recovery (Figure 2.15).

Cyclone Pam was exceptionally strong, but the people of Vanuatu are accustomed to these powerful storms. On 1 April 2020, Tropical Cyclone Harold ripped through the northern islands, having severely affected the Solomon Islands. It would leave Vanuatu, to track towards Fiji and then Tonga. While the rest of the world was understandably distracted by the COVID-19 virus, these Pacific nations responded to widespread destruction without any international presence, maintaining strict control over movement and importation.

This sub-chapter examines some of the challenges that face built environment professionals – predominantly engineers and architects – who work in humanitarian housing. The sector, which goes by the somewhat inadequate name of 'shelter', supports the re-establishment of housing and communities after catastrophic events such as storms, earthquakes and floods, as well as the needs of communities affected and often displaced by conflict. Shelter support can vary from the immediate emergency need for tarpaulins and blankets as protection against the elements, to the procurement of permanent housing.

Figure 2.15 Destruction and recovery in Tanna, Vanuatu, after Cyclone Pam, 2015. (Source: CARE International UK (Bill Flinn))

The discussion in this sub-chapter draws on the extensive experience of the CARE International Emergency Shelter Team as well as a multidisciplinary team that has been researching the wider impacts of good shelter programming and, in particular, the concept of self-recovery.

The humanitarian shelter sector

The International Federation of the Red Cross and Red Crescent Societies (IFRC) and the United Nations Refugee Agency co-lead the coordination group known as the Global Shelter Cluster (GSC, 2021). The former directs the cluster in response to natural hazards; the latter for conflict and mass displacement. The cluster system (there are 11 clusters, including education, health, water and sanitation) was first activated in 2005 in a concerted effort to improve the coordination, effectiveness and equity of humanitarian aid. National shelter clusters are now in place in many hazard-prone countries across the world, or they are set up following a major crisis, and they are the institutional bodies charged with working with governments to set the strategies and technical standards for humanitarian responses.

Although devastating for the small island of Tanna in Vanuatu, Cyclone Pam destroyed or damaged a relatively low number of houses: approximately 17 000 in total compared with over a million in the Philippines after Typhoon Haiyan in 2013. The scale of destruction can defy imagination, and, not surprisingly, mega-events such as Typhoon Haiyan or the 2015 earthquakes in Nepal dominate media and donor attention at the expense of underfunded 'forgotten' disasters. Nonetheless, for each family that loses a home, regardless of it being the result of a headline-grabbing disaster such as Haiyan, or the result of seasonal flooding in Africa that gets barely a mention, the loss is everything that constitutes a home and not just four walls and a roof.

Humanitarian responses, particularly to disasters caused by natural hazards, have tended to follow a phased pattern of immediate emergency response, followed by support to 'early recovery', then sometimes reconstruction, but always with a thread of preparedness for the next event, known as disaster risk reduction (DRR). Increasingly, this is being challenged by an approach that focuses on people's right to choose and set their own priorities and timeframe – in other words, their right to agency – and that recognises that rebuilding a home is a developmental as well as humanitarian imperative.

What is self-recovery? And who decides on the definition of a good house?

Aid agencies are not mass housing providers. There are many disasters that destroy countless houses. Typhoon Haiyan in the Philippines was but one. The earthquakes of Nepal and Haiti in 2015 and 2010, respectively, repeat floods and storms in Africa, and Hurricane Katrina in 2005 are other well-known examples. We now know that the humanitarian sector rarely reaches more than 25% of the total housing need, and frequently that percentage is in single figures (Miranda Morel, 2018; Parrack et al., 2014). One way or another, using their own resources, the remaining 75–90% recover on their own: they self-recover.

Conventionally, aid organisations have adopted a targeted approach, selecting some of the most vulnerable as recipients of shelters, shelter kits or other forms of support that help these families back to a decent home. Technical training and the building of, for example, model houses can extend this support more widely, but nonetheless the coverage is limited. The majority will rebuild, employing whatever means they have available. However, this will often incorporate the same pre-disaster vulnerabilities that caused so much destruction and loss in the first place; and miss the opportunity to rebuild safer, healthier and better.

Recent research conducted in Nepal and the Philippines (Schofield *et al.*, 2019; Twigg *et al.*, 2017) has shown, on the one hand, that self-recovery is an almost inevitable process but also, on the other, that the context in which it happens varies enormously. Generalisations are very few and far between in the humanitarian shelter sector, but most would agree that 'people affected by disaster are not victims; they are the first responders … and the most critical partners' (Jha *et al.*, 2010). A second truism is that no two disasters are ever the same, and understanding the context is of paramount importance. While a fishing family in a coastal community in the Philippines can rebuild their bamboo dwelling within days, the two- or three-storey stone house of a rural family in the mountains of Nepal, by contrast, might take years to complete. Moreover, the consequence of structural failure is quite different in each case: the bamboo house is unlikely to cause harm in a storm that comes with some warning; the Nepali masonry house can be fatal in an earthquake.

Two principles underlie the notion of supporting self-recovery (Flinn, 2019). The first is the near inevitability of the process: people are never passive – recovery starts the day after a disaster, and families and communities will recover even without any support at all from the international community. As a Filipino participant in a recent training event said, 'Self-recovery is what we do anyway' (CARE, 2019). By supporting self-recovery, we are assisting a process that is already under way and one that has been gathering momentum long before the 'experts' arrive. We are relatively small players in a much bigger recovery process.

The second principle is the primacy of people's control, agency and choice. Proponents of self-recovery insist that families and communities have a right to choose their own recovery pathways, prioritising actions according to perceived needs and timeframes (Box 2.1). This may well mean that the choices people make – prioritising their livelihood, for example, over a safely engineered house – could be unpalatable to the practitioner or the donor.

A poor family, already in debt before a crisis, is faced with a series of almost impossible competing priorities: it might be a roof over their heads; the children's continuing schooling; the agricultural cycle; the regeneration of their livelihood. Putting food on the table is likely to outweigh the need for a seismically safe or storm-resistant home.

The home itself also faces a series of competing priorities. The sector is fond of saying that we must 'build back better', but who decides on the definition of 'better'? Structural safety to protect from prevailing hazards is certainly one aspect. However, this must be weighed against size, healthiness, a location for home-based employment, and the prestige and position that the home signals to the neighbours. For a family living illegally in an informal settlement, structural safety is unlikely to be the major concern: with the threat of eviction looming, saving money on the home in order to secure a livelihood will be more important (Flinn and Schofield, in press). For a large rural family with secure tenure, a house with several rooms and an attractive street frontage can be the priority. There is no universal definition of 'better' (Flinn, 2020).

Box 2.1 Recognising self-recovery as a positive force for recovery

The November 2013 Typhoon Haiyan (known locally as Yolanda) was one of the strongest storms ever to make landfall. It swept across a swathe of the Philippines, leaving an extraordinary wake of destruction. Tragedy hit the provincial city of Tacloban, where a two-metre storm surge killed an estimated 10 000 people. The residents do not count the number of destroyed homes, they simply refer to the situation as a washout, with debris scattered far and wide and only concrete floor slabs remaining.

CARE Philippines, working with several local community partners, had a shelter programme that explicitly focused on a self-recovery approach. Almost 16 000 houses were built, and each one of them was unique because families rebuilt according to their own needs and priorities. CARE and its partners supported them with a very small quantity of cash (about £45 per household), roofing materials and technical advice. The programme won the Global South World Habitat Award in 2017.

Judita is a widow and mother of twelve. She took refuge in the school during the night of the storm. As the roof began to be ripped off, everyone formed a human chain and moved from classroom to classroom, holding tightly to each other for safety. The next day Judita returned to her house but found the bamboo wreckage strewn across the street. As a senior citizen, Judita was prioritised by her community self-help group (known in the Philippines as bayanihan*) and her house was rebuilt first. The group built the timber frame and roof with the help of one of her sons. For the rest she hired a carpenter (Figure 2.16).*

(Adapted from Flinn and Echegaray, 2016)

Figure 2.16 Judita's home, Philippines, 2016. (Source CARE International UK (Bill Flinn))

The shelter sector is also beginning to question its singular attention on structural safety. A recent report by InterAction (an alliance of international NGOs) highlights the need for more research into and evidence on the wider impacts of shelter programming. Among other factors, the report identifies physical and mental health, livelihoods, DRR, education, and water and sanitation (WaSH) as sectors that are directly affected by the quality of housing (InterAction, 2020). In particular, there has been a realisation that many housing-related illnesses such as malaria, tuberculosis and diarrhoeal diseases are responsible for far more death and suffering that the combined effects of earthquakes, storms and floods. Poor indoor air quality is a major cause of the 2.5 million deaths each year from lower respiratory infections, primarily pneumonia (Institute for Health Metrics and Evaluation, 2021). By comparison, deaths from natural hazards, although they can vary dramatically year by year, are 60 000 on average (Riche and Roser, 2019). It is not surprising that, at the time of writing, the COVID-19 pandemic has helped to highlight the close correlation between health and housing quality. Overcrowding in refugee camps and post-disaster shelter are now likely to be heightened considerations.

What is the role of engineering?

To adapt the quotation commonly attributed to Voltaire, 'perfect is the enemy of good', is safe the enemy of safer? For engineers, especially those educated in the global north, to advocate for anything less than safe, or less than an agreed code, is anathema. In the UK and across Europe, almost all buildings are constructed to a defined code or standard, sometimes complying with deemed-to-satisfy requirements, but frequently fully engineered by a qualified structural engineer. However, the majority of domestic buildings across the world, and all those in informal settlements, are non-engineered. These are the homes that are repeatedly destroyed by storms, floods, earthquakes and fire.

That everyone deserves a safe place to live is undeniable; adequate housing is a human right (OHCHR, 2014). However, there is an uncomfortable truth: the humanitarian shelter sector does not have the capacity, and nor do many national governments, to ensure that every house is rebuilt after a disaster to a code-compliant level of safety, even if such a code exists for that country. A second uncomfortable truth is that the 80% who self-recover simply cannot afford to rebuild to such codes.

In many ways this discussion is a variant of the quality versus quantity debate. How can we achieve the greatest good for the greatest number? Building safely, for just a few families, does not achieve this. Improving the safety of a house in a fairly modest way – by going from unsafe to safer – can be realised at reasonable cost. Taking the next step and creating a house that is code compliant and 'safe' is much more costly.

Examples may best illustrate the point. Following Global Shelter Cluster guidance, the Filipino fisherman might include some cross-bracing and improved connections into his timber and bamboo house at very little extra cost. It will not be able to withstand the 250 km/h winds of a super-typhoon, but when there is a warning of one of the less severe typhoons that batter the Philippines every year, he will evacuate his family, and, when he returns, his house, with luck, will have sustained no more than minor damage. He feels 'safe enough'. To build a steel-framed building to fully resist the infrequent super-typhoons would be prohibitively expensive.

The remote Himalayan farmer faces very different dilemmas, but the principle is similar. The return period of earthquakes of this magnitude in the Himalayas is in the region of 80 years, but no one knows when the next one will be. Learning and implementing improved walling techniques, such as the inclusion of through stones and risers (stones that bind the wall together), costs nothing; adding a small amount of cement to the mud mortar costs relatively little. Both will result in a stronger wall. Insisting on a reinforced concrete ring beam is expensive and, given poor quality control and deterioration over time, of uncertain efficacy.

Families face all manner of risks. It can be useful to think of these in three different categories. First, are the acute, or 'intensive', risks; these are high-severity, mid- to low-frequency events such as earthquakes and severe storms that can cause untold damage, injury and death. Unexpected landslides and fires fall into this grouping, as might slum clearance and forced relocation. The second category can be called 'extensive risk', and includes frequent and repeat events such as seasonal flooding, monsoon rains, regular storms or volcanic eruptions, and cascade events when one event will trigger another. Finally, there is endemic risk. This is the risk that poor families live with all the time, it is the day-to-day risk. Many of these risks are clearly related to housing quality in its widest sense. Health risks have already been mentioned, and are responsible for hundreds of thousands of premature deaths each year. However, there are several other endemic risks, and many of them are little understood and under-researched by the humanitarian sector. Examples include mental health and well-being (Triveno and Nielsen, 2020), domestic violence and protection (InterAction, 2020). These risks have a disproportionate impact on women and girls, who frequently spend more time in the house than men and boys (CARE, 2016).

Beyond the risks, there are also opportunities. Some of these may be the inverse of the risks: we can improve health and increase protection through encouraging better housing. Others, such as livelihoods for example, are best seen as ways in which good shelter programming can facilitate recovery on many fronts. This can be through the establishment of home-based enterprises such as small convenience stores, backyard cultivation, sewing or taking in laundry (Flinn and Echegaray, 2016). It can also be through construction training or micro-grants to set up, for example, carpentry workshops. Once again, the opportunities can be extensive for women (Kellett and Tipple, 2000).

And what is the role of the engineer?

How do engineers pick their way through this minefield of competing priorities? Much will go against their formal training in meeting standards of safety, complying with codes and their duty of care to society. Of course, they are not alone – all built environment professionals, architects, surveyors and planners come up against this same reality check. The education system will not generally equip engineers for the non-standard, informal, hybrid nature of much of the world's low-economy housing. Where is the computer programme that can meaningfully analyse a timber house, with minimal foundations, woven-bamboo walling and a thatch roof? Understandably, and mindful of the professional insurance that may or may not be valid, engineers are reluctant to propound a judgement on the safety or otherwise of this unfamiliar building type.

In a programme that is supporting the self-recovery process, information and communication are key components. If we accept the primacy of the household's right to decide on, and control, their own recovery pathway and priorities, then our central role becomes that of provider of the best possible information. Frequently this is in the form of 'key messages' that point to simple construction techniques – cross-bracing, improved connections and good foundations for example – that will result in a strong house. These messages are often the result of many weeks of debate before a consensus is found, and are never ranked in order of importance or value for money. Moreover, they inevitably focus solely on structural safety with no reference to the wider impacts of shelter programming. How do you advise a family on the relative merits of, for example, mosquito netting and cross-bracing, or rainwater harvesting and a ring beam?

A few factors do help the engineer to navigate these seemingly impossible questions. The first is the inevitability of the self-recovery process. It is probable that we will be latecomers to the scene and will encounter a process that is already under way, and that has its own pace and rhythm. Families will not all be doing the same thing, as this will depend on their own identified priorities, resources and land. Our role is not to change the direction of travel but to work *with* families and communities to support them along their own recovery pathways and to encourage the construction of safer, healthier and better homes. People need access to good technical information that allows them to make informed choices. This is much more a process of listening and accompaniment than one of rules, codes and diktat.

Secondly, the engineer needs to understand what is 'fit for purpose' and the meaning of 'good enough'. Post-disaster housing, across most of the world, is an incremental process, necessarily evolving from perhaps a temporary shack to something more formal and eventually permanent. At no stage during this process will the home correspond to the engineer's idea of a properly designed house built to known engineering parameters. By encouraging affordable, replicable and simple good construction and good design principles, then the home can be safer, stronger, healthier and better; families can adapt their home to suit their needs and can afford to conduct maintenance (CRS, 2018). This is not to say that families affected by disaster can be fobbed off with something that is merely 'good enough'. Rather, it acknowledges the inevitability of a world of huge need and scarce funding and resources where it is both equitable, appropriate and most effective to abide by the maxim of the greatest good for the greatest number. The alternative is to create 'pockets of excellence', or off-the-shelf houses, for just a fortunate few, which is not fair and cannot be replicated.

It should, at this point, be stressed that what may be true of housing is certainly not so for public community buildings such as schools and clinics. The building of a home must compete with the day-to-day needs of the family's budget and so may not meet the normally acceptable levels of safety. However, there is a duty of care with public buildings, including public housing, that demands a quite different level of engineering scrutiny. In many circumstances these are also the buildings that are used as refuges during storms, and as collective temporary accommodation in the immediate aftermath. However, for self-built housing there is a real question of what is an acceptable level of safety: what can be deemed 'fit for purpose' and 'good enough'?

A shelter programme that supports self-recovery can involve both direct and indirect assistance. Cash grants are used across the humanitarian sector, and they have the capacity to allow choice and agency. Self-recovery projects frequently take a three-pronged approach of cash, some materials and technical assistance. These can all be considered as directly supporting families and communities. The engineer has a role in each aspect: determining the appropriate size of the cash grant and appropriate conditioning of its use (if, indeed, any conditions are appropriate); selecting hard-to-access materials, such as corrugated galvanised iron sheets or hurricane straps, that will make a significant difference to safety or quality; training local builders and raising awareness among families through technical assistance.

Indirect support is less tangible. Families and communities face a series of barriers or obstacles that can hinder their self-recovery process. These fall into many categories. The engineer might be involved in consideration of infrastructure such as the availability of water or access to remote markets. There may be issues of land tenure and disputes with neighbours and within families. Access to finance will almost always be a major concern, and families are vulnerable to loan sharks and overwhelming debt burdens. These are all obstacles to recovery, and lifting these barriers is a very positive, though less visible, form of indirect support. Moreover, many of them will have 'co-benefits', or, in other words, be examples of 'win-wins'. Facilitating access to micro-loans or setting up village-level savings schemes, for example, may help families establish small livelihood projects, as well as support home reconstruction. Improving transport links will help to bring in building materials; but it can also benefit the transport of agricultural goods to market.

It can be seen from this very complex and varied picture presented by the aftermath of a disaster that understanding its context is of paramount importance. This is much more than just an analysis of the needs or an assessment of building damage; it is also more than just a snapshot in time. It requires a participative approach with the affected population that explores their hopes and fears, priorities and likely timelines. It will be a shifting picture that changes as priorities adjust and circumstances alter. There is a need to be adaptable: a project cannot be set in stone.

Conclusions

From the discussion above, it will be clear that the practice of post-disaster, and post-conflict, shelter is multidisciplinary (Twigg et al., 2017). The shelter practitioner cannot cover the breadth of expertise that is needed. As we consider the very wide impact that good (or bad) housing has on safety, livelihoods, physical and mental health, and well-being, it becomes apparent that it cannot be left in the hands of any one built environment professional, be they engineer, architect or planner. However, in reality, the shelter practitioner is often multi-hatted, sometimes inexperienced and frequently flown in, on a short-term deployment, from a different country and culture.

Reducing vulnerability and increasing resilience are part of the current humanitarian discussion. However, the humanitarian engineer, flying into a disaster zone, will be witness to extraordinary resilience in the face of terrible trauma. Moreover, the most severe vulnerabilities – and perhaps the most evident is the exposure to extreme weather events due to accelerating climate change – are the consequence of irresponsible activity in the industrialised world. Supporting families and communities to recover means working closely with them and always seeking to listen and learn, as well as inform. This can only be achieved by working through local partners who know the area, speak the language and have the trust of the population. They are truly the experts; and those of us who have worked across many countries are always humbled by their skills, just as we are humbled by the resilience of the populations we work with.

REFERENCES

ASCE (2021) Code of ethics. https://www.asce.org/code-of-ethics/ (accessed 12/04/2021).

B2P (2014) *Bridges to Prosperity Suspended Bridges Manual*, 4th edn. B2P, Denver, CO, USA. https://www.bridgestoprosperity.org/bridgebuildermanual/ (accessed 12/04/2021).

Bird S and Simon TP (2013) The challenges of humanitarian access bridges in Democratic Republic of Congo. *Proceedings of the Institution of Civil Engineers – Bridge Engineering* **166(2)**: 1–12, 10.1680/bren.12.00002.

CARE (2016) *Gender and Shelter: Good Programming Guidelines*. CARE International UK, London, UK.

CARE (2019) *Soaring High: Self-recovery Through the Eyes of Local Actors*. CARE International UK, London, UK.

CHS Alliance (2014) *Core Humanitarian Standard on Quality and Accountability*. CHS Alliance, Group URD and the Sphere Project, Geneva, Switzerland.

CRS (Catholic Relief Services) (2018) *Extending Impact Study. A Practical Review*. Catholic Relief Services, Baltimore, USA.

Flinn B (2019) *Humanitarian Shelter and the Ethics of Self-recovery: A Discussion Paper*. CARE International UK, London, UK. https://insights.careinternational.org.uk/publications/humanitarian-shelter-and-the-ethics-of-self-recovery-a-discussion-paper (accessed 12/04/2021).

Flinn B (2020) Defining 'better' better: why building back better means more than structural safety. *Journal of Humanitarian Affairs* **2(1)**: 34–43, 10.7227/JHA.032.

Flinn B and Echegaray M (2016) *Stories of Recovery, CARE Philippines Post Haiyan/Yolanda Shelter Response*. CARE International UK, London, UK.

Flinn B and Schofield H (in press) Stay or leave? How recognising self-recovery can support the agency and choice of typhoon survivors in urban Tacloban, Philippines. In *Rethinking Urban Risk and Relocation* (Johnson C. (ed.)). UCL Press, London, UK.

Government of Nepal (2015) *NBC 203. Guidelines for earthquake resistant building construction of low strength masonry*. Government of Nepal, Ministry of Physical Planning and Works, Department of Urban Development and Building Construction, Kathmandu, Nepal.

GSC (2021) https://www.sheltercluster.org/global (accessed 12/04/2021).

ICC (International Code Council) (2018) *International Building Code 2018*. International Code Council, Washington, DC, USA.

ICE (Institution of Civil Engineers) (2016) *Civil Engineering Procedure*, 7th edn. ICE, London, UK.

Institute for Health Metrics and Evaluation (2021) Global Burden of Disease (GBD). http://www.healthdata.org/gbd (accessed 12/04/2021).

InterAction (2020) *The Wider Impacts of Humanitarian Shelter and Settlements Assistance*. InterAction, Washington, DC, USA.

IStructE (Institution Of Structural Engineers) (2006) *Standard Method of Detailing Structural Reinforcement*, 3rd edn. IStructE, London, UK

Jha A, Barenstein JD, Phelps PM, Pittet D and Sena S (2010) *Safer Homes, Stronger Communities*. World Bank and Global Facility for Disaster Reduction and Recovery, Washington DC, USA.

Kellett P and Tipple G (2000) The home as workplace: a study of income-generating activities within the domestic setting. *Environment and Urbanization*, **12(1)**: 203–213, 10.1177/095624780001200115.

Medair (2021) Who we are. https://www.medair.org/who-we-are/ (accessed 12/04/2021).

Miranda Morel L (2018) *Shelter Assistance: Gaps in the Evidence*. CARE International UK, London, UK. https://insights.careinternational.org.uk/media/k2/attachments/Shelter-assistance_gaps-in-the-evidence_Discussion-Paper_November-2018.pdf (accessed 12/04/2021).

OCHA (UN Office for the Coordination of Humanitarian Affairs) (2015) *Vanuatu: Tropical Cyclone Pam. Situation Report No. 9*. OCHA Regional Office for the Pacific, Geneva, Switzerland. https://reliefweb.int/sites/reliefweb.int/files/resources/OCHA_VUT_TCPam_Sitrep9_20150323.pdf (accessed 12/04/2021).

OHCHR (Office of the UN High Commissioner for Human Rights) (2014) *The Right to Adequate Housing*. OHCHR, Geneva, Switzerland. https://www.ohchr.org/documents/publications/fs21_rev_1_housing_en.pdf (accessed 12/04/2021).

Parrack C, Flinn B and Passey M (2014) Getting the message across for safer self-recovery in post-disaster shelter. *Open House International* **39(3)**: 47–58, 10.1108/OHI-03-2014-B0006.

RedR (2021) *Our history.* https://www.redr.org.uk/About/Our-History (accessed 12/04/2021).

Riche H and Roser M (2019) Causes of death. Our World in Data. https://ourworldindata.org/causes-of-death (accessed 12/04/2021).

Schofield H and Flinn B (2018) People First: Agency, choice and empowerment in the support of self-recovery. In *The State of Humanitarian Shelter and Settlements* (Sanderson D and Sharma A (eds)). International Federation of Red Cross and Red Crescent Societies, Geneva, Switzerland, pp. 29–34.

Schofield H, Lovell E, Flinn B and Twigg J (2019) Barriers to urban self-recovery in Philippines and Nepal. Lessons for humanitarian policy and practice. *Journal of the British Academy* **7(s2)**: 83–107, 10.5871/jba/007s2.083.

Sphere (2018) *The Sphere Handbook: Humanitarian Charter Minimum Standards in Humanitarian Response*, 2018 edn. Sphere, Geneva, Switzerland.

Triveno L and Nielsen O (2020) Home sane home. World Bank Blogs. https://blogs.worldbank.org/sustainablecities/home-sane-home (accessed 12/04/2021).

TRL (Transport Research Laboratory) (2000) *A Design Manual for Small Bridges: Overseas Road Note 9*, 2nd edn. TRL, Crowthorne, UK.

Twigg J, Lovell E, Schofield H *et al.* (2017) *Self-recovery from Disaster: An Interdisciplinary Perspective*. Overseas Development Institute, London, UK.

UK Highways Agency (2001) *Design Manual for Roads and Bridges,* vol. 1: *Highway Structures: Approval Procedures,* section 3: *General Design,* part 14, BD 37/01: *Loads for Highway Bridges.* 1(3), Part 14, BD 37/01. Highways England, Birmingham, UK

World Bank (2021) The World Bank in DRC. https://www.worldbank.org/en/country/drc/overview (accessed 12/04/2021).

Georgia Kremmyda
ISBN 978-0-7277-6468-3
https://doi.org/10.1680/hce.64683.055

Chapter 3
Humanitarian engineering framework and practices

Abstract

The chapter highlights the particularities of the 'wicked' and unpredictable problems that humanitarian engineers are called to manage. Engineers are called to reject solutions with minimal social content or solely technical content. The chapter provides guiding principles when considering a framework for the appraisal, design, planning, execution and sustainability of humanitarian engineering projects. It reviews the steps and issues involved in the execution and management of humanitarian projects towards the selection and recommendation of a transformative suite of approaches to humanitarian challenges such as clean water and sanitation, shelter provision, affordable and clean energy, sustainable cities and communities or the bigger issues of poverty and inequality, urbanisation land management and issues of social context. Case studies, lessons to be learnt and successes from over 20 years of projects delivered from Engineers for Overseas Development are given, providing an overview of humanitarian engineering practices.

BEING TRANSFORMATIVE

Regan Potangaroa

Introduction

As humanitarian engineers we want to help (EHRID, 2018). But what sort of help and who we help are the immediate issues. Even when we do know the 'who', we still have difficulty with the 'what'. So, we ask those affected. Often, they are uncertain or cannot guess what that could be. We realise it is hard to even get stuff to them and that there are additional issues of quality and cost. And that any help needs to be sustainable and preferably locally sourced. We finally realise that help has to be cross-discipline but nonetheless 'simple'. This is the context of humanitarian engineering, which starts with the simple desire to help (The Nation, 2013).

The problem

But all or many of the answers are pitched at an organisational level and not for the practitioner above who just wants to help.

For example, the Transformative Agenda (IASC, 2005) was a programme within the humanitarian sector marked by an urgency to bring about positive change. It represented a renewed commitment by humanitarian actors to work together in an accountable manner to achieve collective results. The goals were to strengthen and streamline humanitarian responses, especially for declared Level 3 (IASC, 2012a) emergencies, by the following protocols

- *Protocol 1. Humanitarian System-Wide Scale-Up Activation: Definition and Procedures* (replacing *Humanitarian System-Wide Emergency Activation: Definition and Procedures*) (IASC, 2012b)
- *Protocol 2. 'Empowered Leadership' in a Humanitarian System-Wide Scale-Up Activation* (replacing *Concept Paper on 'Empowered Leadership'* – revised in March 2014 and again in 2018) (IASC, 2014, 2018)
- *Responding to Level 3 Emergencies: What 'Empowered Leadership' Looks Like in Practice* (IASC, 2013a)
- *Reference Module for Cluster Coordination at Country Level* (revised July 2015) (IASC, 2015a)
- *Reference Module for the Implementation of the Humanitarian Programme Cycle, Version 2.0* (July 2015) (IASC, 2015b)
- *Accountability to Affected Populations: The Operational Framework* (IASC, 2013b)
- *Inter-Agency Rapid Response Mechanism (IARRM) Concept Note* (December 2013) (IASC, 2013c)
- *Common Framework for Preparedness (ARP)* (October 2013) (IASC, 2013d)
- *Emergency Response Preparedness (ERP)* (Field Testing, August 2015) (IASC, 2015c)
- *Multi-Sector Initial Rapid Assessment Guidance* (Revision July 2015) (IASC, 2015d).

All or much of this addressed the organisational and procedural requirements of the United Nations (UN)-based humanitarian system (Messina, 2014), although the idea of an empowered leadership in the third point above did strike a chord with practitioners. We will come back to this later. But there is nothing really for the humanitarian engineer who wants to 'help'.

This divide that the humanitarian engineer finds themselves standing at is neatly captured in the International Federation of Red Cross and Red Crescent Societies video 'The shelter effect' (IFRC, 2010). In this video, Ang and Chen's family live off the land from fishing, poultry and growing rice. Every year they experience flooding. But one year they are hit twice. They lose their self-sufficient lifestyle, and their health is compromised. The video then suggests a transformational change for the family by raising their house by two steps, treating the timber posts for water and concreting the floor. Over time those simple changes protect the family and their heath, and enable a rapid reset after further flooding that frees up their income to eventually buy a cow. The benefits continue grow, to include the school and eventually the village. The IFRC called this the 'shelter effect' – a transformational change brought about by raising the house by two steps, water treatment of the timber and concreting the slab. But the video does not answer by who or how this transformational change was determined. What was the empowering leadership or vision for this transformational change?

And hence, how can I be 'transformative' as a humanitarian engineer?

The Logical Framework Approach

The first formal attempt to make sense of humanitarian contexts was the Logical Framework Approach (LFA) (AusAid, 2003). It is a 'linear' objective-based approach that consists of nine sequential steps (Örtengren, 2004)

1. analysis of the project's context
2. stakeholder analysis
3. problem analysis/situation analysis
4. objectives analysis
5. planning of the activities
6. resource planning
7. indicators/measurement of the objectives
8. risk analysis and risk management
9. analysis of the assumptions.

The results of this approach and analysis are communicated by a 4×4 matrix, with different agencies and donors having slightly different formats and syntax (Engineers without Borders, 2015; Institution of Structural Engineers, 2021; Save the Children, 2018).

It was devised and deployed by the US Agency for International Development in the 1970s because aid funding was not being effectively used, as planning was too vague, the management responsibility was unclear, and evaluation was an adversarial process (PCI, 1970).

However, there still remained challenges with the LFA, which included the following (Bakewell and Garbutt, 2005; Jackson, 1997).

- LFAs were not suitable where there was a degree of uncertainty or disagreement about what was the main problem.
- LFAs were seemingly rigid once they were formulated and could not readily be altered as any new reality was realised. Because of this, project managers tended to treat them as an administrative tool and a requirement of the funding agency rather than a useful planning and design management aid. There was minimal 'ownership'.
- Consequently, LFAs did not readily enable monitoring of 'unintended consequences'.
- LFAs limited the emergence of potential solutions, innovative thinking and adaptive management.

Nonetheless, many organisations still use the process.

Tame and wicked problems

Rittel and Webber (1973) stated that 'one of the most intractable problems is that of defining problems (of knowing what distinguishes an observed condition from a desired condition) and of locating problems (finding where in the complex causal network the trouble really lies)'. They were thinking of planning problems such as 'the location of a freeway, the adjustment of a tax rate, the modification of a school curricula or the confrontation of crime'. They postulated that there were two types of problems: one was 'tame' or benign while the other was 'wicked'. A tame problem (Ritchey, 2013) is one that

- has a relatively well-defined and stable problem statement
- has a definite stopping point (i.e. we know when a solution is reached)
- has a solution that can be objectively evaluated as being right or wrong
- belongs to a class of similar problems that can be solved in a similar manner
- has solutions that can be tried and abandoned.

A wicked problem is the opposite of these 'tame' criteria.

Interestingly, wicked problem approaches are becoming more mainstream, as evidenced by their uptake by various government aid agencies (Australian Government, 2007; Ramalingam *et al.*, 2014; Spratt, 2011) and the variety of wicked problem approaches, such as the following.

- Grint (2008) suggested clumsy solutions, which are derived from merging the egalitarian, individualist and hierarchist approaches.
- Conklin (2005) – whose approach is probably the most well known – suggested dialogue mapping, which represents both the social complexity and wicked problem aspects in one seemingly mind map approach, to achieve a shared understanding.
- Branenburger and Nalebuff (1999) suggested changing the 'rules of the game' rather than accepting the 'game' you are given.
- Roberts (2000) suggested (in a similar way to Grint) that power is not dispersed using an authoritarian coping strategy. If it is dispersed that way and it is contested, a competitive strategy is used; otherwise, a collaborative strategy is used. The authoritarian, collaborative and competitive coping strategies appear to be similar to, respectively, the hierarchist, egalitarian and individualist solutions in Grint's analysis.
- Alexander *et al.* (1977) suggested the hierarchical decomposition approach (set theory or network analysis). This breaks a complex network into a family tree, with the parent or root issues at the top of the tree and the many children or dependant issues at the bottom.
- Potangaroa *et al.* (2014) used a quality of life approach to measure community 'well-being' and resilience. The approach seeks to metricate the process and thereby manage it.

The unproblemising suite

Despite all this advice, the operational perspective for a humanitarian engineer remains unclear. How can they find not just a workable solution but one that is transformational?

Based on over 200 deployments to such situations, I would like to offer the following approach, which I term the 'unproblemising suite'. There are currently three key components to the suite

- the new tool spproach
- the unproblemising spproach
- the contra-matrix spproach.

They emerged from the shelter sector because that is the field in which I work, but the three approaches do have a 'universal feel'.

The new tool approach

This approach seeks to directly 'change the game' through the application of a new tool as follows

- use a new tool to redefine the 'issue', or 'change the problem' to 'change the game'
- reflect on what it might be telling you, look for the patterns – what new information is it providing and what changes in approach and strategy would this represent?

New tools that I have used include the following, which are discussed briefly together with a short description of what the tool is, why it is a new and the 'new game' it provided. It should be noted is that most of these tools are not actually new but rather well known – just not within the humanitarian sector.

- Scala penetrometer: a portable 15 kg tool for ascertaining soil-bearing capacity (GroundTest, 2021). It is the *de facto* standard for house building in New Zealand (Standards New Zealand, 2011). However, there are currently no field methods readily available to test soil bearing capacity in the humanitarian field. This is critical for rebuilding in the recovery and reconstruction phases in non-refugee situations and for camps in refugee situations. It allows an engineered approach to less than perfect ground conditions, especially with the foundation design recommendations of Stockwell (1977).
- Space syntax (DepthMap, 2021): a computer-based method used to check the visibility of points within a space. It opens up discussion on the use of space, including the meaning and location of public and private areas, which are often guessed at.
- Drone mapping: a process of producing maps from drone flights over a site. It is new in terms of access provided by the process of the 'drone today, map tomorrow' methodology and in terms of its speed and ease of access to the technology. The production of maps in real time benefits many disciplines, such as health (e.g. maps of where patients are located in a camp), protection (e.g. the provision and effectiveness of night street lighting) and law (e.g. finalising contract documentation for civil works construction).
- Quality of life (Potangaroa *et al.*, 2014): a survey tool that was developed to quantify well-being or happiness. It is adapted from the Depression Anxiety and Stress Scale instrument for the humanitarian context, and has the ability to discern between those who are happy and those who are unhappy in a disaster (which admittedly sounds contradictory). Nonetheless, it provides a way to filter data and arrive at the actual outcomes of affected populations rather than relying solely on the outputs of the process employed.
- Hierarchical decomposition (Alexander, 1963, 1964): a process that has the ability through set theory to break down wicked problems into a tree structure, with the root causes at the top of the tree and consequential actions at the bottom. Thus, the process brings an untapped intelligence to the issues it is applied to. This is discussed later in more detail.

■ Virtual reality (VR) and the active shooter (Wikipedia, 2021): VR together with three-dimensional (3D) scanning produces an immersive and unlocking environment that provides access to insights previously not available. In the active shooter (US Department of Homeland Security, 2008) scenario, a shooter is released autonomously into the 3D scanned version of a virtual office, and the active shooter protocols of run, hide and fight are enabled. The new knowledge for the humanitarian engineer is knowing when to move between each of these three protocols.

The value of these tools is their ability to include those affected in the analysis – not just to consult but to partner and co-design.

The unproblemising approach

This approach seeks to address the problem through a sequencing process that whittles away until one arrives at the actual problem. It can be iterative by repeating the same process, but need not be. The trick is to keep your eye on the problem. This is perhaps the most used and known of the three approaches. The steps are as follows

(*a*) Develop technical resolutions for the apparent need(s) of the users.
(*b*) Review the problem and iterate until at least three other needs are resolved, then calculate the costs and process to implement. Are there further 'opportunities'?
(*c*) Has 'better' in terms of those affected been achieved?

The reflective phase in (*b*) can invariably be missed in an emergency, and the process is a workflow that loops back to (*a*) with no (*c*), as in the case study below.

Case study: the Rohingya emergency response disaster drone

Rohingya Muslims from the northern areas of Rakhine State in Myanmar moved to Bangladesh in response to extreme violence by the Myanmar military. Starting on 25 August 2017 and continuing for 3 months, 745 000 refugees streamed into the Cox's Bazar area of Bangladesh. This was the fourth and largest influx compared with previous ones in 1978, 1991 and 2016. Refugees initially settled in Kutupalong and Nayapara, and in other areas in the Teknaf and Ukhia districts. The largest of these settlements was Kutupalong Camp (KTP), which became the fourth largest city in Bangladesh. The camp consisted of rolling sand dunes 20–30 m in height that at its highest elevation was 70 m above sea level.

The land was cleared of vegetation and then terraced for tent sites. By November 2017 the major influx had completed, and rationalisation of KTP and camps others started, with a focus on KTP as the largest refugee city in the world. The camp area of KTP was small, and only $10\,m^2$ per person rather than the Sphere requirement of 30–$45\,m^2$ per person (Sphere, 2018) could be achieved. As a comparison, the population density of KTP was twice that of Dhaka – which is often itself touted as a crowded city. It was not good.

The UN High Commissioner for Refugees (UNHCR) and the International Organisation for Migration (IOM), which were responsible for one-half of the camp each, set about lobbying the Bangladesh Government for additional land. Some land was provided, and decongesting and relocation options were implemented. However, the Bangladesh Government had made its own plans, and built on a sandbar in the Bay of Bengal called Bashan Char (Banerjee, 2020; DW News, 2019). This drew a negative response from the humanitarian community, and efforts to keep refugees on land in Cox's Bazar were prioritised.

But as the monsoon season approached, it became evident that clearing of vegetation and terracing were the two worse things to do for landslides to occur. And it was clear that plans to address flooding had to be completed. When risk maps for these were completed, they showed that 23 330 people would be affected by landslides and 85 876 people would be affected by flooding. And if both happened, then they would impact 75 985 people, as some would unfortunately be affected by both landslides and flooding.

However, another problem emerged from the risk study, namely that 222 133 people would be on islands and thus inaccessible. Because of its low elevation, ponding would be present for months and, furthermore, the water would be contaminated from the washed-out and collapsed latrine and sewerage infrastructure. A scenario-planning exercise was instigated, with government response, non-governmental organisation (NGO), medical, IOM and UNHCR teams, together with representatives of the refugees, to see what could be done. It became clear that any response would have to be based in KTP, as the coastal highway from Cox's Bazar to KTP would be cut off. But that still left 222 133 people marooned on islands.

The final part of the plan was the development, together with RedR Australia (2021), of a disaster drone – one that could fly in any kind of rain and most wind. The protocol and training for that was then pushed down throughout the UNHCR teams, to ensure that there were people available for deployment from Cox's Bazar to KTP. The strategy for these drones was for them to go out first and establish priorities for assistance across the camps. Until that information was available, no coordination could realistically happen.

This case study underlines the sequencing and the iterative parts of the unproblemising approach. The risk analysis was iterative, while the drone part was sequential. The work flow process was over nearly 1 year, which, in hindsight, should have been reduced. But we did not know about the numbers that would be isolated on islands in KTP. Additionally, there was pressure for further land from the government, and hence a danger of falling into the trap of the 'bigger the figures, the better the case'. And this was balanced against the negative idea of shifting refugees out to Bhasan Char island. The tension was real, and it was paramount to retain academic integrity to produce the above risk numbers.

Only one life was lost due to flooding or landslides, which we took as a success (UNHCR, 2018).

Contra-matrix approach
This approach seeks to address the problem through the identification of contradictions between two or more project requirements, referred to as the contra-matrix. This is perhaps the least known or used approach from the suite. The steps are

(*a*) identify and set up the contra-matrix
(*b*) identify the area(s) of contradiction
(*c*) solve by iteration, typically using solutions from outside the contra-matrix.

It was well after this work that I came upon TRIZ – a Russian acronym for 'theory of inventive problem-solving' (MindTools, 2021; TRIZ Journal, 2020). This approach to invention seeks out such contradictions and then applies 39 other contradictory states to develop what is termed (interestingly) a contradiction matrix. The first of the three principles of TRIZ is that all contradictions have their answers outside the domain applicable to a situation. The second principle is that an invention happens when a contradiction is resolved. The third principle is the existence of the 39 other states for contradictions. It is clear that the TRIZ approach can offer pathways for addressing such contradictions.

Case study: 2016 Vunikavikaloa Arya Primary School, Fiji – Tropical Cyclone Winston

Tropical Cyclone Winston was a category 5 event that hit Fiji on 20 and 21 February 2016. At its peak, the cyclone was estimated by the Fiji Meteorological Service to have 10 min of sustained winds of 64 m/s (230 kph), gusting to peaks of 90 m/s (325 kph), making it the most severe cyclone to land in the South Pacific (ReliefWeb, 2016a). The Fijian Government estimated that 350 000 people would be impacted, with later records showing 41 deaths and 131 people injured. Among the extensive damage were 229 schools (Fiji Government, 2016). These 229 schools (Ashna Kumar, 2016) would play a key role Fiji's recovery, and hence were prioritised by the Fijian Government.

The Fiji Institute of Engineers (FIE), working closely with the Department of National Planning within the Ministry of Finance, set up an adopt-a-school programme based on an assessment and cost to rebuild the schools done by volunteer Fijian Registered Engineers. The programme targeted foreign government aid, multilateral organisations, donor organisations and corporations with the ability to undertake major rebuilding or repairs (ReliefWeb, 2016b; RNZ, 2016).

This (especially the costing) was an innovative idea, and it allowed the Fijian Red Cross Society (FRCS) supported by the New Zealand Red Cross (NZRC) to adopt Vunikavikaloa Arya School in Ra Province. It is the collector school for five villages where the inhabitants are ethnic Indian and predominantly cane cutters. The cyclone had lifted the roof off the school, and consequently all school materials inside were lost. And hence the engineering work would involve replacing the roofs on the main block, the toilet block and the on-site teacher accommodation.

However, the position of the FRCS was that the rebuild would not stop at the physical structure but instead would start at that point. Its approach was to make the rebuild a learning and healing experience for the community. It meant that volunteers from the surrounding five villages would need to be involved in the rebuilding, because strengthening the community was as important as strengthening the building. Moreover, the FRCS saw the school project as the start of an ongoing relationship rather than any conclusion to it. This desire then set up the contra-matrix condition as follows.

First, the NZRC had issues regarding the extent of the damage and the design of the school, in addition to issues of community engagement. They also had questions about the timeline and the costs for the work, and how it would be implemented: it had to be achievable using a volunteer team with minimal to no construction experience.

The Ministry of Education (MoE) had standards and codes that needed to be adhered to, and it was not completely clear what these might be but it was anticipated they would include door and window details, surface finishes and an overall professional finish. There was also the question of whether the school would or should be an evacuation centre, which would further add to the details list.

The Construction Implementation Unit (CIU) – the Fijian Government department overseeing the construction of all schools – needed to know that the adoption programme for Vunikavikaloa Arya would be completed on time and to accepted standards, as mentioned above. Consequently, it was expecting a comparable level of quality of finished construction to their other professional contractors. It aimed to have all 220 schools completed by the school opening date (end of January 2017). However, the FRCS proposal had a team with minimal to no experience.

The FIE was involved in ensuring that the school design was signed off by a Fijian Registered Engineer. There was also discussion among the engineering professionals about revising the wind design speed (Rattan and Sharma, 2005; Structural Engineering Blog, 2020). Fijian codes current at the time were gazetted in 1990 and were based on an earlier Australian wind code (AS 1170:1989). The Fijian wind loading was based on wind region C for Australia but based on a category 4 cyclone. However, the FIE was promoting design to category 5 cyclone levels (taken as 66 m/s for a 3 s wind gust) with somewhere between 75 and 77 m/s for evacuation centres. It was problematic to get a categorical answer for the previous 3 s wind gust design speed for a category 4 cyclone used in Fiji prior to Cyclone Winston, but it seemed to be around 58 m/s. Thus, the change to category 5 represented a 29% increase in load while the evacuation centre would increase by 76%. These were significant increases and represented a (needed) step change in building design, but also represented a huge barrier to the FRCS proposal.

The teaching staff at the school were desperately looking forward to their classrooms being back and operational. The schools' tents worked well, but were difficult and hot to teach in, small and problematic for pinning up material. In addition, they needed to be regularly shifted because the ground gets soft and muddy in the rainy season. They were concerned that the school roll had dropped from around 250 to 230 students, and were keen to get their damaged school back fully functioning.

Moreover, the students too were keen to get back to normal (Vunikavikaloa Arya School, 2017).

Finally, the building context in Fiji was characterised by a shortage of materials such as steel sections for the roof (needed for the proposed increased wind loads), the additional costs for such materials, and a shortfall of skilled tradespeople to fabricate and install structures given that there were 220 schools of concern. Material was being imported but the process was slow; and even when it arrived there was still the issue of too few tradespeople. Vunikavikaloa Arya Primary School was one of 220 schools all in the same situation, with it having the additional requirement of community engagement.

The core of the contra-matrix was the desire to use an inexperienced team on a project that was above normal industry and sector requirements for category 5 cyclone loading rather than category 4.

Strategy

Besides the core contra-matrix there were several associated ones, of skill and material shortages, timelines, costs and cost escalation, plus a need to provide quality control for the MoE and the CIU. But the core presents the problem to be solved. There were no quick ways to train people to levels that are beyond the standards of the industry. So, the physical needs of the project were assessed. Making the steel sizes smaller and more closely spaced addressed the issue of material availability and the scale of work – but perhaps not the cost. Taking out the welding requirements was a significant change, as these were fundamental to the skill and cost issues. Issues of 'right and wrong' were also included in this assessment, especially those that would prioritised Vunikavikaloa over the other 219 schools overseen by the CIU – the do no harm principle (OECD, 2010), localisation (Birzer and Hamilton, 2019; ReliefWeb, 2016c) and the 'commitment to strengthening the independence of local and national leaders in humanitarian action and decision-making, in order to better address the needs of affected populations' (Council for International Development, 2018).

The strategy that materialised from this thinking had two parts, namely

- The use of light-gauge steel sections that could easily be screwed together to form trusses. This addressed quality control and the shortage of skilled tradespeople while contributing to cost savings and maintaining timelines.
- Construction of the roof on the ground and lifting it into place in sections. This reduced the risk of having volunteers high up in the air and minimised the need for the school to find and use additional tents while the roofing was constructed.

This need for a strategy had been anticipated, and discussions had been initiated with FrameCad NZ (FrameCad, 2021). It had the capacity, the software, the machines and the training programmes. Moreover, it had a 20 ft container that could produce the steel sections on site. Hence, it was suggested that FrameCad be appointed as a preferred subcontractor for this project. There were two other light-steel suppliers in Suva, but they were not able to specifically fabricate and design building components or train construction staff. Neither were they willing to pass on their technology to local institutions.

The FrameCad approach computerised the roof truss design for both analysis, design and fabrication. Thus, the design wind speed could be directly dialled into the programmed designs produced for checking by a Fijian Registered Engineer. The steel was in coils, and thus the finish and thickness could be controlled. The FrameCad fabricating machines produced the various truss sections, which were joined with screw fixings in predetermined hole locations. This produced a light-weight structure that could readily be fabricated by a semi-skilled team and lifted into place. So, it brought something in from outside the construction field.

Most of the material in Fiji such as roofing is sourced from either Australia or New Zealand, so doing this did not seem to run contrary to the idea of localisation. Hence, as part of its tender FrameCad was asked to include the roofing. This made the construction simpler, as all the screws, bolts, connectors and roofing arrived together with the folding machine.

It was interesting to note that out of the 220 schools only 40 were ready for the new year of teaching, and Vunikavikaloa Arya Primary School was one of them. All of the other schools used large steel sections that had to be imported and needed skills in terms of welding, fabrication and erection that were uncommon to the area. A video of the construction process is available (Fiji Red Cross Society, 2017).

Conclusions

Humanitarian engineers are often viewed as technical problem-solvers by the rest of the humanitarian sector. And, surprisingly, we probably also see ourselves that way.

The problems we are trained on are 'tame', with well-trodden solution paths and usually one answer. In the field the problems we are called upon to manage are 'wicked' and unpredictable, with more than one answer. They are very different from the classroom or training room. The unproblemising suite may assist that transition. Ultimately, the academic and technical content of courses must extend out and engage with the ground reality that awaits the profession, in addition to the humanitarian engineer.

We need to reject the idea of minimal social content to our solutions, and also the idea of being solely technical. The unproblemising suite adopts and adapts several approaches in a search for the key problem. It uses processes such as 'changing the game' and being a 'problem-maker' rather than a 'problem-taker', but all are based on what is encountered in the field.

The unproblemising suite means that humanitarian engineers can be more effective and focused on the goal of humanitarian engineering to bring about transformational change. The three methods in the suite are applicable to others areas covered in this book, such as clean water and sanitation, affordable and clean energy, industry innovation and infrastructure, and sustainable cities and communities, or the bigger issues of poverty and inequality, urbanisation land management and social context. How you approach these problems will determine what alternatives you will be able to offer. And hence the unproblemising suite.

At the end of the day, we as humanitarian engineers seek to help people; and perhaps the joy of the work we do is that we are able to help a lot of people.

THE HUMANITARIAN WORK OF ENGINEERS FOR OVERSEAS DEVELOPMENT (EFOD)

Ian Flower and Sally Sudworth

Introduction

For over 20 years EFOD has been delivering humanitarian engineering developments in sub-Saharan Africa, encouraging young civil engineers and others from the construction industry to enhance their career development (EFOD, 2021). Volunteers working in small teams have delivered 30 individual schemes in remote parts of Africa, including school buildings, health centres, mills and grain stores, boreholes and latrines. Thousands in communities of the rural poor have benefited from its work, while hundreds of EFOD engineers have benefited from the experience of delivering all aspects of a scheme, including design, fundraising, construction supervision and commissioning.

Genesis

EFOD started in February 2000 in response to a challenge given during an Institution of Civil Engineers (ICE) Wales lecture in Cardiff. Colin McCubbin, working as a civil engineer for Mercy Ships, gave a presentation on the provision of simple sanitation in Africa. He had just returned from The Gambia, and invited members to provide latrines in downtown areas of the capital Banjol. Six recent civil engineering graduates volunteered, researched local standards and designed simple latrines. ICE members raised funds for construction, the volunteers raised funds for their travel and accommodation. They took leave from their employment, and visited the site in pairs for just 2 weeks. Working with Banjol Council engineering staff, they hired labour, purchased materials, supervised construction and commissioned a pair of four-cubicle latrine blocks for the local authority, both to programme and budget over a period of 10 weeks.

Local communities had access to quality latrines for the first time, and local labour benefited from the employment and the training provided by the EFOD volunteers. In turn, the volunteers benefited from the delivery of the first complete scheme of their careers, developing skills in project management, while learning building techniques from the workers.

Growth

In 2000, Voluntary Service Overseas and others were offering opportunities for travel and humanitarian aid, but there were few opportunities for engineering graduates in full-time employment to use their skills in humanitarian aid. The Register of Engineers for Disaster Relief (RedR) required a commitment of 3 months or more, and Engineers Without Borders arrived in the UK the following year, offering opportunities for undergraduates. EFOD was and still is a valuable opportunity for volunteers to gain excellent experience in the delivery of humanitarian relief schemes for some of the world's poorest folk. Besides, all six original volunteers wanted to repeat the exercise, and sound principles had been established, so EFOD was established.

Over the years EFOD has grown in size, with new groups forming to meet the enthusiasm of recent graduates and technicians to deliver humanitarian aid, initially as a subgroup of ICE Wales Cymru. It registered as a charitable company in 2011 as groups formed in Bristol, Birmingham, Manchester, London and, finally, South West Wales. The operating principle of each group remains the same, designing a project that delivers aid to remote communities in The Gambia, Uganda and Ghana. The value of the schemes has increased significantly from £6000 for the first latrines up to £110 000 for medical facilities in north-east Uganda. Volunteers are encouraged to participate in all aspects of the scheme, raising funds, design, managing both the programme and budget, and visiting the site in pairs to gain site experience, essential for civil engineers training, when they share their skills with the local workers.

Over 1000 civil and other engineers together with technicians and apprentices have contributed to EFOD's success. Over £1 million has been raised and around 400 visits made to site, generally for 2 weeks although some have remained for 1, 3 and 8 months.

Client selection

Although construction work has been completed in The Gambia and Ghana, and design work in Ethiopia and Zambia, the majority of projects continue to be delivered in north-east Uganda, resulting in a regular trained workforce, which is available to work with EFOD volunteers. This has made project identification, delivery and monitoring much easier.

Careful client and project selection is necessary to ensure the objectives of a scheme are fully understood and the solution appropriate, and will be utilised and maintained. Continued contact with the client following completion, including regular evaluations, have proved invaluable.

EFOD's status as a charitable company limits the client base to charities, NGOs, community benefit organisations and local government bodies. Medical incinerators, latrines and water schemes have been delivered for hospitals, grain stores and mills for women's cooperatives, and boreholes and latrines for communities.

Clients have included UK-based charities the SaltPeter Trust, Act 4 Africa, Give a Child a Hope, Care and Share Foundation, Dolen Ffermio and Teso Education Student Support (TESS), all working in Uganda.

Funding

This is often one of the most challenging aspects, particularly for large schemes in excess of £30 000, where construction has generally progressed in stages as funds are raised. Ideally, schemes are fully funded before site work begins, providing continuity of employment for the labour force of up to 40. It is also beneficial for the supervisory team, since it allows the new pair to meet at handover and be briefed by the departing pair. On large schemes, work has progressed in stages, and photographs of early site work are used to support fundraising initiatives and grant applications.

Most of the large grants have been provided by major engineering consultancies and contractors. Mott MacDonald, Arup and AECOM have been very generous over 20 years, appreciating the value of the training on offer for the staff involved in EFOD schemes. Travel grants have come from the Wales for Africa Fund provided by the Welsh Government as educational support for participants from Wales.

Each volunteer is required to fund part of their insurance, travel and subsistence costs. Group fundraising schemes have included many sponsored cycle rides, raffles, formal dinners and more innovative ideas such as sponsored head shaving. The regional branches of ICE local to each group have also offered great support.

Health and safety and well-being

Health and safety training on site has been necessary during EFOD supervision. UK volunteers are well schooled in sound health, safety and welfare (HS&W) practices, and share this with the local workforce. Poor early practices such as concrete mixing in bare feet have been replaced with adequate tools and personal protective equipment, and regular tool box talks are given to ensure the adoption of sound practices. All temporary works must be fabricated on site since specialist equipment is not available in the remote areas. Safe and appropriate working-at-height practices have been developed. There is little mechanical lifting equipment available, so scheme design has to be appropriate for lifting by the workforce.

Of equal importance is the HS&W of the volunteers. EFOD uses a travel agency and a travel insurer familiar with aid work, and approved drivers in country to ensure safe travel. Accommodation is chosen with care, usually following use by directors and senior volunteers.

Sustainability

Sustainability has been at the heart of EFOD projects. Where possible, an on-site water supply is provided at the onset of each scheme, ensuring there is water to build, leaving a supply for the local community once work is complete. Rainwater harvesting has been provided on most of the schemes, and, in recent years, solar power for lighting and phone charging within scheme buildings.

Interlocking stabilised soil blocks (ISSBs) have been cast on site on all schemes since 2008, avoiding the need to fell trees to fire clay bricks, and reducing cement in block laying because of their regular shape. On recent schemes, up to 95% of materials have been sourced within 15 km of the scheme, to reduce the carbon footprint.

In the last year, 4000 tree seeds have been sown and nurtured, so they can be planted around existing and new schemes as carbon offset of flights and travel and the scheme itself.

Case studies

Medical waste incinerators, Uganda, 2002–2013

In 2002 EFOD Cardiff was challenged to build an incinerator for Soroti Regional Referral Hospital in north-east Uganda. The 1950s structure was inoperable, and blood tubes from the wards were scattered around, at a time when estimates suggested 25% of the population was HIV positive. A team of six EFOD engineers researched the problem, and opted to introduce a De Montfort-type incinerator based on a design by D. J. Picken. A total of £6000 plus travel costs had been raised in early 2003, when civil unrest caused by the Lord's Resistance Army prevented safe access to Soroti. The team identified an alternative mission hospital in similar need in Masaka, south-west Uganda, which was unaffected by the problem, and a waste incinerator was constructed by the hospital maintenance team supervised by EFOD over a period of 8 weeks (Figure 3.1).

Peace was restored in Soroti in 2005, and eight engineers travelled out in pairs to build a Picken incinerator at the local regional referral hospital. Volunteers from Bristol formed a new EFOD group led by one of the eight, and went on to provide a third incinerator, this time in

Figure 3.1 Hospital waste incinerator, Masaka, Uganda

Atatur Hospital in 2007, and a new group in Manchester did the same for Kumi Mission Hospital, following approaches from Dolen Ffermio.

The Picken incinerator burns at very high temperatures and so needs repair every 5 years or so. EFOD has repaired the Soroti and Kumi models, and the hospital maintenance team in Masaka did so themselves. The local project manager from the build in Soroti Hospital has been involved in the repairs, and now has a business building and repairing medical waste incinerators, employing some of the original workforce from 2005.

Soroti Baptist Medical Centre, Uganda, 2005–2009

When building the incinerator at Soroti Hospital, EFOD Cardiff had more volunteers willing to travel to site than necessary. In response, SaltPeter Trust provided funds for them to lay the foundations of a 16-room not-for-profit medical centre in the grounds of Soroti Baptist Church at the same time. Welcoming the challenge, the team offered to complete the design, and to construct the superstructure of the £100 000 single-storey building. Work restarted in early 2008 with supervision from 16 EFOD engineers visiting in pairs. A frame was built from reinforced concrete columns cast within hollow blocks, topped with a reinforced concrete ring beam. The infill walls were formed from ISSBs cast on site using murrum won from site excavations. A steel-framed mono-pitched roof was planned, but construction was affected by a steel shortage in early 2009. Electoral disputes in Kenya led to the closure of the border with Uganda, and so the team made a last-minute design change, constructing timber trusses clad with locally produced profiled roof sheeting. Two 50 000 l circular rainwater harvesting tanks were built to the rear of the unit using curved ISSBs.

Figure 3.2 Soroti Baptism Medical Centre, Uganda

Fitting of doors and windows, and internal finishes, were completed before handover to SaltPeter Trust in October 2009. The centre formally opened in May 2010 (Figure 3.2), and currently employs 15 staff, providing low-cost medical and dental care, together with laboratory services, to the residents of Soroti and the surrounding districts. The centre benefits from passive ventilation, and this is so successful that the nurses complain they have to wear cardigans in the mornings!

SaltPeter Trust is managed by a civil engineer, teacher, doctor and dentist, who maintain regular contact with the centre, providing medical and management support. Teams of volunteers visit once or twice a year to provide additional medical assistance and training, and help with regular maintenance and improvements. The centre is making a real difference in Soroti, providing low-cost medical, dental and laboratory services to the general public and over 2000 children from seven local child development centres. A grant from the Welsh Government financed a series of pop-up clinics in nearby villages in 2019–2020, providing free treatment for over 3000 patients. The centre also hosts a project for children with cancer, funding their travel to specialist centres in Kampala.

Soroti District continues to benefit from the services offered by the medical centre, thanks to its provision by young civil engineers and architects working for EFOD.

Care & Share Soroti Orphanage (CASSO), Uganda, 2009–2010
South East Wales-based charity Care & Share asked EFOD to assist with the development of an orphanage near Soroti for 83 young street children. CASSO provided most of the funding, and EFOD Cardiff moved on to the design of CASSO once the Baptist Medical Centre design

was complete. Construction of four accommodation blocks, a large hall and a kitchen followed, all benefiting from passive ventilation. ISSBs were used for the columns and walls, supporting timber trussed roofs. EcoSan toilet blocks were introduced for the first time by EFOD, based on a Ugandan Ministry of Education guide. Solids and fluids are separated, the former decay in underground chambers for 1–2 years before removal and being used as manure, while the latter are diluted and used to water crops. Two latrine blocks with showers were constructed, and their use was a success. Water could not be found beneath the site, so mains water was brought in, with some rainwater harvesting from roofs. However, CASSO found difficulty funding both the construction and support for all the orphans. The orphanage ran for 6 years, accommodating 40 of the children, who were educated at local schools. When the orphanage closed, the children were sent to boarding schools. The buildings stood empty for 3 years, suffering from vandalism, before acquisition by a very successful school in Soroti. Now, 240 nursery and primary children are being educated on the site, and additional classrooms have been added. Interestingly, 25 of the pupils are orphans and live with house parents in two of the original accommodation blocks. The new owners were not familiar with EcoSan toilets, and had planned to demolish and replace them with pit latrines. However, following a recent visit from EFOD representatives, they have agreed to put them back into operation, which demonstrates the value of maintaining contact with schemes, and providing advice on maintenance.

Grinding mills and grain stores, Uganda, 2009–2020

Koutulai mill and grain store

SaltPeter Trust had been promoting agricultural support for 3 Widows Cooperatives in Kachumbala, Uganda, and one of the groups in Koutulai requested help with the provision of a grinding mill (Figure 3.3) to assist the 30 members with milling and storage of their crops. EFOD West Midlands started in 2009, led by a member from EFOD Bristol, and adopted this

Figure 3.3 Koutulai grinding mill, Uganda

as its first scheme. A three-roomed single-storey structure with a mill, store and office was built by ten graduate civil engineers, introducing ISSB construction to the remote area 5 km from the main highway. Completion of the mill provided free milling for the members, and a grinding business yielding profit for the cooperative.

The following year, EFOD West Midlands funded a borehole on the site, and returned to construct a block of six EcoSan toilet cubicles for the cooperative and community.

In 2011 the same group was joined by two architects, and designed and constructed a village hall to seat 150 people, together with three adjacent offices, once again using ISSBs, and a timber trussed roof providing passive ventilation.

The following year saw the addition of a translucent sheeted patio, to double the capacity for weddings and events, and in 2013 a 900-bag grain store was built, and the site fenced to add security, all by EFOD West Midlands and local labour.

The cooperative has grown from 30 to 86 in number. The hall is being used for community training sessions, by both the cooperative and the local authority, and by a church on Sundays. A vocational school started 2 years ago, providing tuition in tailoring, computing, building and driving for local residents.

Grain storage and milling significantly increase the value of grain for cooperative members. A large-scale maize storage trial in 2018 was very successful. Weevils can seriously affect stored maize, reducing its weight in spite of regular exposure to the sun to drive them away. Subsequent trials storing maize in Growpro bags (hessian sacks lined with polythene) have proved successful, with no weight loss, and an improved storage business will continue. However, similar trials with cassava have failed, and further storage trials are necessary to identify simple longer-term storage techniques.

In 2019, Zukuka Bora Coffee, an Mbale-based charitable company, hired space to store green coffee from the lower slopes of the nearby mountain, prior to export, and has developed an EFOD funded drying unit on the site, providing employment for 6 local residents.

Construction of the Koutulai scheme has led to the development of a trading centre on adjacent land, to the benefit of the local community.

Nyakoi Mam Riang grain store

The success at Koutulai has been repeated in nearby Nyakoi village (EFOD, 2014). EFOD North West, which built the incinerator at Kumi Hospital, moved on to provide a 480-bag grain store for the Mam Riang (Empty Field) Cooperative (Figure 3.4), a group of 64 married women. It was built by a local workforce, supervised by 16 engineers over 16 weeks in 2014. Water was available from a nearby borehole, and ISSBs were cast on site from locally arising murrum, and used to construct the single-storey building. To comply with good storage practices, the dual pitched roof had a 'top hat' for ventilation, and occasional translucent panels. Solar panels were added to power lighting and electrical outlets.

Figure 3.4 Nyakoi grain store, Uganda

Kachumbala mill and grain store

In 2011, an apprenticeship scheme based in Coleg Sir Gâr in Ammanford, Wales, requested involvement in EFOD for its construction apprentices. EFOD South West Wales formed, managed by staff from the college and local building companies, and embarked on a scheme to build a mill and grain store for a third cooperative of 30 widows, based in the town of Kachumbala.

Designed by EFOD Cardiff, a group of 14 apprentices delivered another 900-bag structure, following the provision of a borehole providing water for both the building and the community.

Use of the three grain stores has been beneficial to all the cooperative members, although they have yet to be fully filled. Evaluations by EFOD directors in 2013 and 2017 confirmed that the lot of each of the members has improved considerably. All are better fed, better dressed and have better housing, and several have been able to fund tertiary education for their children.

TESS Vocational School, Uganda, 2011–2013

TESS had plans for a new centre in Kapir, and EFOD Cardiff worked with its Spain-based architect, providing civil and structural design for offices and accommodation blocks, together with a rainwater-harvesting scheme. Phase 1 was constructed by EFOD, hiring and supervising a local labour force. Phase 2 was carried out by a local contractor with occasional EFOD inspections.

Kpone Saduase Sewing School, Ghana, 2011–2013

Cambridge Education, a part of Mott MacDonald, invited EFOD Bristol to build a sewing school in Kpone Saduase for the widows of the village to teach skills and provide work. EFOD Bristol designed a building to resist earthquake activity, providing a workshop/training room and accommodation for visiting tutors, and managed the construction with visits to site over a period of 2 years.

Malera borehole refurbishment, 2012–2015

In 2012, EFOD Cardiff was asked by the community of Malera, Uganda, to refurbish a borehole for a community with large herds of cattle (Figure 3.5). Rock head in the area is high, and droughts and floods are common, contributing to poverty in the area. Over 5000 cattle were lost in 2007 during a serious drought.

EFOD Cardiff accepted the challenge to improve the water supply for the area, and surveyed 13 existing holes, before opting to refurbish a deep borehole in Kabole, 1 km from the town. Pumping tests suggested it was capable of delivering 6000 l per hour. In the 1960s the borehole had benefited from a diesel pump and two large storage tanks, discharging into cattle troughs. Civil unrest had destroyed the engine, and corrosion the steel tanks. A hand pump had been installed, limiting the yield to a maximum of 400 l per hour. The consequences were long queues from the community, and dry troughs for the cattle.

Figure 3.5 Malera borehole refurbishment, Malera, Uganda

A solar-powered turbine pump delivering water into two 20 000 l HDPE tanks was the preferred solution. During site visits, EFOD employed local staff and laid tank bases and a store within a fenced enclosure, before commissioning Kampala-based Innovation Africa to install pipework, solar panels and the pump. Finally, the two cattle troughs were refurbished. The scheme was commissioned in August 2015, and provides water for 4000 people through a bank of drop taps, and 5000 cattle via the troughs, vastly improving the conditions for villagers in the area.

Management and the collection of fees has proved difficult. The borehole committee has changed three times, sub-county boundaries have moved twice, and the borehole is now under the control of another new sub-county. Fiscal management is necessary to ensure the caretaker can be paid and funds accrued for regular maintenance, and EFOD has regular meetings with the community, encouraging families to contribute 45p per month, and farmers 22p per head of cattle per year, to provide a fund for maintenance.

Kathy's Centre for Women and Children, Uganda, 2015–2017

Act for Africa, a Manchester-based charity founded by Kathy Smedley, requested support from EFOD North West for the construction of a centre to accommodate a crèche, meeting room and medical laboratory for vulnerable women and children of Mayuge (Figure 3.6). EFOD volunteers responded with a scouting trip, establishing the need, identifying suppliers of materials, and appropriate accommodation, before designing a well-ventilated building complete with solar power and standard and disabled EcoSan toilets. This was built by a local labour force, supervised by 14 EFOD volunteers over a period of 2 years. ISSB manufacture was introduced to the area, and the press gifted to Act 4 Africa, to provide employment for the workers trained during the scheme.

Figure 3.6 Kathy's Centre for Women and Children, Mayuge, Uganda

Kathy's Centre now provides the community with vital health and support services, pre-school education, and agricultural, enterprise and financial training.

Kachumbala Health Centre 3 Maternity Unit, Uganda, 2014–2017

During construction of the Kachumbala Mill, the First Minister of Wales made a visit to the site, and was entertained by members of the Widow's Cooperative. The chair of the local health board and Local Council V asked for help to improve maternity provision for the area. A new maternity unit was built by EFOD South West Wales. More details are given in Chapter 4.

Kumi Hospital pump house and latrine projects, Uganda, 2013 to date

Dolen Ffermio, a charity based in Mid Wales commissioned EFOD to build a medical waste incinerator for Kumi Mission Hospital in 2012. While there, it became clear that the hospital water supply from the nearby lake was occasionally threatened when the lake was in flood. An old electrically driven pump house and its power supply had been destroyed during civil unrest, and a new pump house had been constructed in 2004, but at a level prone to flooding during very high lake levels. A bund had been built around it to prevent inundation, and hand pumping was required every 6 h to prevent submersion of two diesel-powered pumps. EFOD London started work in 2013, designing and constructing a simple solution, raising one of the pumps by 1.65 m and reconnecting the pipework. The task was undertaken by the hospital maintenance team under EFOD supervision, but the pipework had to be fabricated in Kampala. The water supply has been secured, although a second pump needs to be lifted to increase the resilience of the water supply.

A repair of the 2012 waste incinerator was completed in 2018, and a new bank of 16 batteries provided to restore emergency power to the operating theatres.

Current schemes in Uganda

EFOD London is building EcoSan latrines and showers for patients at Kumi Hospital.

EFOD Cardiff is developing Faith Nursery and Primary School for 300 children of Kwarikwar near Kachumbala.

EFOD North West is developing a coffee facility for Zukuka Bora Coffee, part of Jenga charity in Mbale.

EFOD West Midlands is developing a mill and grain store in Matugga, near Kampla, for the Revival Centre, a school, orphanage and women's centre supported by the charity Give a Child a Hope from York in the UK.

Company structure

EFOD is a small company registered with both Companies House and the Charities Commission in the UK. There are no staff, nor premises, all involved are volunteers and it has a turnover in excess of £100 000 each year. Overheads are carefully controlled, and the major overhead is the cost of company insurances.

The two authors of this sub-chapter are directors of EFOD, supported by Julian Howe and Jane Hodgson, all with significant experience in the civil engineering industry. Each oversees one of the EFOD groups and is supported by a group mentor, usually a former EFOD project manager. In turn, each group has a project manager, treasurer, secretary, and fundraising, publicity and design leaders. Dave Cousins, another civil engineer reviews all EFOD designs.

Lessons learnt

EFOD has had many successes, but many lessons have been learned over the course of 20 years.

Offer schemes rather than await requests?

In 2001, during a meeting with the leader of Mbale Council in Uganda, one of the authors offered to build latrines in the town, a repeat of The Gambia scheme. The offer was accepted, land allocated at the showground, and two blocks of five cubicles were completed by six EFOD Cardiff engineers. However, the site was remote, the leader lost his seat in the end-of-year elections, and the council failed to take responsibility for the security and maintenance of the latrines. All subsequent schemes have been at the request of a trusted body, with care taken to ensure a facility will be used as intended and monitored to ensure continued use and maintenance.

Change the EFOD format?

On one scheme, EFOD was asked to work with outside advisors, and supervise a contractor rather than hire labour in the normal way. The scheme was completed, but the programme and budget over ran. The conclusion was to beware of unnecessary changes to a successful format.

Keeping control of the finances?

On a few schemes the client has asked to put all funds through their own books. This has worked in the main, but can lead to delays in payment for materials and labour, and result in delays on site. On one scheme the value of a cheque was increased, resulting in missing funds, and the scheme was stopped for some months until the loss was repaid.

On another scheme, money transferred by Western Union was intercepted by thieves who had hacked an email account and withdrawn it using forged documents. Attempts to recover proved time consuming and costly. Court orders were necessary to obtain details from the bank that released the funds, the police required travel costs to Kampala to investigate, several reports were necessary, and the UK Fraud Line and Charities Commission were notified. All attempts to recover the stolen funds proved futile when the bank was foreclosed by the Government of Uganda.

Subsequently, notification of funds transferred by Western Union was sent in two parts, some details by email, the rest as a text message, with instructions to withdraw on the same day. Schemes are now funded using cash cards for smaller sums, and, for larger funds, by trusted charity accounts in the country of the scheme, where EFOD can attend the bank with the account signatory and withdraw money.

EFOD engineers are accountable for all the scheme funds provided, and are required to maintain daily accounts supported by receipts for materials, travel and accommodation, payment logs for the labour costs, and end of scheme accounts.

Successes

Since its inception in 2000 EFOD has successfully delivered a series of projects that provide humanitarian aid for rural communities in sub-Saharan Africa, providing excellent experience early in their careers for EFOD volunteers.

Thirty infrastructure projects have been completed and are being maintained, to the benefit of several thousand people. The facilities provided continue to both supply services and create employment for the rural poor. Over £1 million has been raised to fund the costs of the

Table 3.1 Vision and mission positions of selected engineering-based INGO's

Organisation	Vision/mission
RedR	A world in which sufficient competent and committed personnel are available and responding to humanitarian needs. We build the knowledge and skills of individuals and organisations for more effective humanitarian action. (RedR UK, 2021)
Engineers Without Borders	Our Vision is for a world where 'everyone has the access to the engineering leadership and capability required to lead a life of opportunity and be free from poverty in all its forms'.Our Mission is to 'connect, educate and empower people through humanitarian engineering'. Humanitarian engineering uses a people-centred, strengths-based approach to improve community health, well-being and opportunity. Our mission statement explains how we plan to achieve our vision. (Engineers Without Borders, 2021)
Engineers Against Poverty	Our vision is that all people have access to adequate, affordable and resilient infrastructure enabling a world free of poverty. Our mission is to promote infrastructure policy and practice with sustainable social, economic and environmental impacts that contribute towards the elimination of poverty. (Engineers Against Poverty, 2021)
Engineers for a Sustainable World	We strive to empower students and professionals to tackle sustainability challenges. We work towards a sustainable world supported by a network of passionate engineers. (Engineers for a Sustainable World, 2021)
Engineering for Change	E4C's mission is to prepare, educate and activate the international engineering workforce to improve the quality of life of underserved communities around the world. We do this by providing resources and platforms that accelerate the development of impactful solutions and ensure public health and safety around the globe. (Engineering for Change, 2021)

purchase of land, materials, labour, flights and subsistence. Training in management and construction has been provided freely, and many of the local construction workers have developed additional skills, some going on to set up their own businesses.

Over 600 civil and mechanical engineers and technicians, architects, environmentalists, scientists and building apprentices have joined EFOD for a period in the early stages of their careers, designing and delivering projects that have varied in cost from £6000 to £110 000. Hundreds more have assisted with fundraising activities. They have made around 400 overseas visits, generally for 2 weeks in pairs, to work with a local workforce, and gained valuable management and construction experience in the process. Most have gone on to receive professional qualifications based in part on the work they have done with EFOD. Some of the apprentices have changed direction and completed degree courses following their time in Africa. Engineers have done the same, moving from contracting to consulting, others have opted to work overseas for some years following their experience with EFOD.

EFOD has been presented with several awards from the construction industry, including ICE, the Association of Consultancy and Engineering, and Constructing Excellence. But the most significant reward has been the knowledge that very many lives have been improved by the humanitarian aid provided.

Conclusions

Civil engineers and other professionals in the construction industry in the early years of their careers, with support from senior colleagues, are capable of using their skills to deliver humanitarian aid, providing significant support to some of the poorest in sub-Saharan Africa, while enhancing their own training and experience.

REFERENCES

Alexander C (1963) *HIDECS 3: Four Computer Programs for the Hierarchical Decomposition of Systems which Have an Associated Linear Graph*. Department of Civil Engineering, Massachusetts Institute of Technology, Cambridge, MA, USA.

Alexander C (1964) *Notes on the Synthesis of Form*. Harvard University Press, Cambridge, MA, USA.

Alexander C, Ishikawa S, Silverstein M, Jacobson M, Fiksdahl-King I and Angel S (1977) *A Pattern Language*. Oxford Press, Oxford, UK.

Ashna Kumar S (2016) Big push to complete the schools rebuild outlined. *Fiji Sun*, 25 Sep. https://fijisun.com.fj/2016/09/25/big-push-to-complete-the-schools-rebuild-outlined/ (accessed 12/04/2021).

AusAid (2003) *AusGUIDElines: 1. The Logical Framework Approach*, updated 20 June 2003. Australian Agency for International Development, Canberra, Australia.

Australian Government (2007) *Tackling Wicked Problems: A Public Policy Perspective*. Commonwealth of Australia, Canberra, Australia.

Bakewell O and Garbutt A (2005) *The Use and Abuse of the Logical Framework Approach*. Swedish International Development Cooperation Agency, Stockholm, Sweden.

Banerjee S (2020) *From Cox's Bazar to Bhasan Char: An Assessment of Bangladesh's Relocation Plan for Rohingya Refugees. ORF Issue Brief 357*. Observer Research Foundation, New Delhi, India. https://www.orfonline.org/research/from-coxs-bazar-to-bhasan-char-an-assessment-of-bangladeshs-relocation-plan-for-rohingya-refugees-65784/ (accessed 12/04/2021).

Birzer C and Hamilton J (2019) Humanitarian engineering education fieldwork and the risk of doing more harm than good. *Australasian Journal of Engineering Education* **24(2)**: 51–60, 10.1080/22054952.2019.1693123.

Brandenburger A and Nalebuff B (1999) The right game: use game theory to shape. *Strategy Harvard Business Review* **76(7)**: 67–105.

Conklin J (2005) *Dialogue Mapping: Building Shared Understanding of Wicked Problem*. Wiley Press, Chichester, UK.

Council for International Development (2018) *Localisation: Implications for the International Non-government-Organisation Sector*. Council for International Development, Wellington, New Zealand. https://www.cid.org.nz/connect/news/localisation-implications-for-the-international-non-government-organisation-sector/ (accessed 12/04/2021).

DepthMap (2021) DepthmapX v0.8.0. https://github.com/SpaceGroupUCL/depthmapX/releases (accessed 12/04/2021).

DW News (2019) Bangladesh plans to relocate Rohingyas to 'refugee island'. *https://youtu.be/6FvaBN1ClZ8 (accessed 12/04/2021).*

EFOD (2014) EFOD NW – Grain for Gain 2014. https://youtu.be/oRyipvY2Q5c (accessed 12/04/2021).

EFOD (2021) https://efod.org.uk (accessed 12/04/2021).

EHRID (Engineers for Humanitarian Relief and International Development) (2018) *Global Design Sprint – October 2017*. BuroHappold Engineering, Bath, UK. http://www.happoldfoundation.org/wp-content/uploads/2018/02/Global-Design-Sprint-_-final_web.pdf (accessed 12/04/2021).

Engineering for Change (2021) Who we are. https://www.engineeringforchange.org/who-we-are/ (accessed 12/04/2021).

Engineers Against Poverty (2021) Our mission and vision. http://engineersagainstpoverty.org/about-us/our-mission-and-vision/ (accessed 12/04/2021).

Engineers for a Sustainable World (2021) About. https://www.eswglobal.org/faq (accessed 12/04/2021).

Engineers without Borders (2015) How to develop a logical framework (Logframe). https://youtu.be/OHEPVS2TAHk (accessed 12/04/2021).

Engineers without Borders (2021) About us. https://www.ewb.org.nz/about_us (accessed 12/04/2021).

Fiji Government (2016) Adopt a school. For our children's sake. For our nation's future. http://www.adoptaschool.gov.fj (accessed 12/04/2021).

Fiji Red Cross Society (2017) Facebook post 29/07/2017. https://fr-fr.facebook.com/FijiRedCrossSociety/videos/students-at-vunikavikaloa-arya-school-in-ra-province-have-been-studying-in-tents/1589665994379983/ (accessed 08/02/2021).

FrameCad (2021) http://framecad.com (accessed 12/04/2021).

Grint K (2008) Wicked problems and clumsy solutions: the role of leadership. *Clinical Leader* **1(2)**: 54–68.

GroundTest (2021) Scala penetrometer kits and accessories. http://www.groundtest.co.nz/scala-penetrometers.html (accessed 12/04/2021).

IASC (Interagency Standing Committee) (2005) IASC Transformative Agenda. https://interagencystandingcommittee.org/iasc-transformative-agenda (accessed 12/04/2021).

IASC (2012a) *How the System Responds to L3 Emergencies*. IASC, Geneva, Switzerland. https://reliefweb.int/report/world/how-system-responds-l3-emergencies (accessed 12/04/2021).

IASC (2012b) *Humanitarian System-Wide Scale-Up Activation*. https://interagencystandingcommittee.org/humanitarian-system-wide-scale-activation (accessed 12/04/2021).

IASC (2013a) *Responding to Level 3 Emergencies: What 'Empowered Leadership' Looks Like in Practice*. IASC, Geneva, Switzerland. https://interagencystandingcommittee.org/iasc-transformative-agenda/documents-public/3-what-empowered-leadership-looks-practice (accessed 12/04/2021).

IASC (2013b) *Accountability to Affected Persons: The Operational Framework*. IASC, Geneva, Switzerland. https://interagencystandingcommittee.org/accountability-affected-people/documents-public/operational-framework-accountability-affected (accessed 12/04/2021).

IASC (2013c) *Inter-Agency Rapid Response Mechanism (IARRM)*. IASC, Geneva, Switzerland. https://interagencystandingcommittee.org/working-group/documents-iasc/inter-agency-rapid-response-mechanism-iarrm-0 (accessed 12/04/2021).

IASC (2013d) *Common Framework for Preparedness*. IASC, Geneva, Switzerland. https://interagencystandingcommittee.org/iasc-reference-group-risk-early-warning-and-preparedness/documents/iasc-common-framework (accessed 12/04/2021).

IASC (2014) *Concept Paper on 'Empowered Leadership'*. IASC, Geneva, Switzerland. https://interagencystandingcommittee.org/iasc-transformative-agenda/documents-public/concept-paper-empowered-leadership-revised-march-2014 (accessed 12/04/2021).

IASC (2015a) *Reference Module for Cluster Coordination at Country Level*. IASC, Geneva, Switzerland. https://interagencystandingcommittee.org/iasc-transformative-agenda/iasc-reference-module-cluster-coordination-country-level-revised-2015 (accessed 12/04/2021).

IASC (2015b) *Reference Module for the Implementation of the Humanitarian Programme Cycle, Version 2.0*. IASC, Geneva, Switzerland. https://interagencystandingcommittee.org/iasc-transformative-agenda/iasc-reference-module-implementation-humanitarian-programme-cycle-2015 (accessed 12/04/2021).

IASC (2015c) *Emergency Response Preparedness (ERP)*, draft for field testing. IASC, Geneva, Switzerland. https://interagencystandingcommittee.org/iasc-transformative-agenda/documents-public/iasc-emergency-response-preparedness-draft-field-testing (accessed 12/04/2021).

IASC (2015d) *Multi-Sector Initial Rapid Assessment Guidance*. IASC, Geneva, Switzerland. https://interagencystandingcommittee.org/iasc-transformative-agenda/documents-public/multi-clustersector-initial-rapid-assessment-mira-manual (accessed 12/04/2021).

IASC (2018) *Protocol 2. 'Empowered Leadership' in a Humanitarian System-Wide Scale-Up Activation*. IASC, Geneva, Switzerland. https://interagencystandingcommittee.org/iasc-transformative-agenda/iasc-protocol-2-empowered-leadership-humanitarian-system-wide-scale (accessed 12/04/2021).

IFRC (2010) The shelter effect. International Federation of Red Cross and Red Crescent Societies. https://youtu.be/Lf2z38u2djA (accessed 12/04/2021).

Institution of Structural Engineers (2021) Working in the humanitarian or development sectors https://www.istructe.org/resources/guidance/working-in-humanitarian-development-sector/ (accessed 12/04/2021).

Jackson B (1997) Designing Projects and Project Evaluations Using The Logical Framework Approach. IUCN Monitoring and Evaluation Initiative.

Messina C (2014) Humanitarian leadership: more than just about leaders. Humanitarian Practice Network. https://odihpn.org/blog/humanitarian-leadership-more-than-just-about-leaders/ (accessed 12/04/2021).

MindTools (2021) TRIZ: a powerful methodology for creative problem solving. https://www.mindtools.com/pages/article/newCT_92.htm (accessed 12/04/2021).

OECD (Organisation for Economic Co-operation and Development) (2010) *Conflict and Fragility: Do no Harm. International Support for State Building*. OECD, Paris, France. https://www.oecd.org/dac/conflict-fragility-resilience/docs/do%20no%20harm.pdf (accessed 12/04/2021).

Örtengren K (2004) *The Logical Framework Approach: A Summary of the Theory Behind the LFA Method*. Swedish International Development Agency, Stockholm, Sweden.

PCI (Practical Concepts Incorporated) (1970) *Guidelines for Teaching Logical Framework Concepts*. PCI, Washington, DC, USA.

Potangaroa R, Santosa H and Wilkinson S (2014) The application of quality of life metrics. In *Disaster Management: Enabling Resilience* (Masys A (ed.)). Springer, London, UK, pp. 227–265.

Ramalingam B, Laric M and Primrose J (2014) *From Best Practice to Best Fit: Understanding and Navigating Wicked Problems in International Development*. Overseas Development Institute, London, UK.

Rattan SP and Sharma RN (2005) Extreme value analysis of Fiji's wind records. *South Pacific Journal of Natural Science* **23(1)**: 1–8, 10.1071/SP05001.

RedR Australia (2021) https://www.redr.org.au (accessed 12/04/2021).

RedR UK (2021) *Our Vision, mission and strategy*. https://www.redr.org.uk/About/Our-vision,-mission-and-strategy (accessed 12/04/2021).

ReliefWeb (2016a) *Fiji: Severe Tropical Cyclone Winston. Situation Report No. 5 (as of 25 February 2016)*. OCHA Regional Office for the Pacific, Geneva, Switzerland. https://reliefweb.int/report/fiji/fiji-severe-tropical-cyclone-winston-situation-report-no-5-25-february-2016 (accessed 12/04/2021).

ReliefWeb (2016b) School's out: Cyclone Winston impacts education. https://reliefweb.int/report/fiji/school-s-out-cyclone-winston-impacts-education (accessed 12/04/2021).

ReliefWeb (2016c) *Five reasons why the 'localisation' agenda has failed in the past - and four reasons why things may now be changing*. https://reliefweb.int/report/world/five-reasons-why-localisation-agenda-has-failed-past-and-four-reasons-why-things-may (accessed 12/04/2021).

Ritchey T (2013) Wicked problems: modelling social messes with morphological analysis. *Acta Morphologica Generalis* **2(1)**: 1–8.

Rittel H and Webber M (1973) Dilemmas in a general theory of planning. *Policy Sciences* **4**: 155–169.

RNZ (2016) Fiji engineers to assess cyclone damage. https://www.rnz.co.nz/international/pacific-news/299156/fiji-engineers-to-assess-cyclone-damage (accessed 12/04/2021).

Roberts NC (2000) Wicked problems and network approaches to resolution. *International Public Management Network Review* **1(1)**.

Save the Children (2018) Humanitarian logframes. https://youtu.be/L8n4PwpFsHc (accessed 12/04/2021).

Sphere (2018) *The Sphere Handbook: Humanitarian Charter Minimum Standards in Humanitarian Response*, 2018 edn. Sphere, Geneva, Switzerland.

Spratt J (2011) Is Development Wicked? Using a Wicked Problems Framework to Examine Development Problems. NZADDs Working Paper.

Standards New Zealand (2011) NZS 3604. Timber framed buildings. https://www.standards.govt.nz/sponsored-standards/building-standards/nzs3604/ (accessed 08/02/2021).

Stockwell M (1977) Determination of allowable bearing pressure under small structures. *New Zealand Engineering* **32(6)**: 132–135.

Structural Engineering Blog (2020) What you need to know about differences in wind-speed reporting for hurricanes. https://seblog.strongtie.com/2017/09/need-know-differences-wind-speed-reporting-hurricanes/ (accessed 12/04/2021).

TRIZ Journal (2020) 40 inventive principles. https://triz-journal.com/40-inventive-principles-examples/ (accessed 12/04/2021).

UNHCR (UN High Commissioner for Refugees) (2018) *Independent Evaluation of UNHCR's Emergency Response to the Rohingya Refugees Influx in Bangladesh August 2017–September 2018. Evaluation Report December 2018*. UNHCR, Geneva, Switzerland. https://www.unhcr.org/5c811b464.pdf (accessed 12/04/2021).

US Department of Homeland Security (2008) *Active Shooter: How to Respond*. US Department of Homeland Security, Washington, DC, USA. https://www.dhs.gov/xlibrary/assets/active_shooter_booklet.pdf (accessed 12/04/2021).

Vunikavikaloa Arya School (2017) Facebook post 31/05/2017. https://www.facebook.com/friendsofVunikaPS/ (accessed 12/04/2021).

Wikipedia (2021) Active shooter. https://en.wikipedia.org/wiki/Active_shooter (accessed 12/04/2021).

Wilentz A (2013) Letter from Haiti: life in the ruins. *The Nation*, 9 Jan. https://www.thenation.com/article/archive/letter-haiti-life-ruins/ (accessed 12/04/2021).

Georgia Kremmyda
ISBN 978-0-7277-6468-3
https://doi.org/10.1680/hce.64683.085
ICE Publishing: All rights reserved

Chapter 4

Learning from humanitarian engineering projects

Abstract

The chapter provides case studies of best practices and challenges reported in humanitarian engineering projects. The case studies are provided by professionals who commend on the availability, accessibility, affordability, and scalability of interventions in the developing and world. The chapter shares knowledge, experiences and good practices on the folowing. How can collaborative working be better supported, including across international boundaries? How might organisations with different missions and roles (e.g. humanitarian relief, development assistance and environmental protection) work more effectively together to build resilience? How might local communities and actors be empowered to make choices about how to build resilience, and what are the constraints on this? Who is excluded or marginalised in the process of building resilience and how might more inclusive participatory processes be developed? How do these challenges differ in different socio-economic and cultural situations? How does building resilience intersect with issues of gender, voice, power and inequality?

DELIVERING A NEW MATERNITY UNIT IN KACHUMBALA, UGANDA

Ian Flower and Dan Flower

Introduction

Engineers for Overseas Development (EFOD) is a small charitable company that manages teams of young civil engineers and others in construction in the delivery of humanitarian aid projects to enhance their management and organisational skills as they work towards professional qualifications. Teams are set a project to design a solution, raise funds, and travel to site in pairs for just 2 weeks to hire labour, buy materials, supervise construction and commission the scheme. More details are given later in this sub-chapter.

EFOD South West Wales (SWW) formed in 2011, to provide an opportunity for apprentices training at Coleg Sir Gâr in Ammanford to work with engineers and technicians delivering schemes in Uganda, supported by several building and construction companies in the area.

The commission

In 2014, EFOD SWW was constructing a grinding mill and grain store in Kachumbala for a widow's cooperative, supported by the UK charity SaltPeter Trust, when Carwyn Jones, then the First Minister for Wales, visited the site to see the work his government had helped to fund. The cooperative members had arranged a party to welcome him with music, singing and dancing. William Wilberforce, then the charismatic leader of Bukedea Council, asked the first minister for support to improve the maternity provision at the nearby health centre for the women of Kachumbala.

The EFOD SWW management team, including Delwyn Jones of TAD Builders and Anthony Rees of the Cyfle Shared Apprentice Scheme, were there with the authors for the visit, together with a team including UK doctors and nurses delivering pop-up clinics in nearby villages. A delegation met the District Health Officer the next day to obtain details of the improvements they sought, and then went on to visit the health centre.

A seven-roomed building had been constructed in 1950, and by 2014 was serving a regional population of 55 000. The maternity ward had just two cramped rooms to accommodate labour, delivery and recovery for mothers from the local community. The ward often exceeded capacity, and the post-delivery recovery time was between 6 and 18 h before discharge, resulting in 40% of women choosing to give birth at home without medical support. The centre was run by a committed clinical officer and 17 support staff, including three midwives. The staff were struggling to provide health care from inadequate accommodation.

The existing health centre (Figure 4.1) is well placed, and sited in a strategic position just north of Kachumbala town, at the interchange between the main Mbale–Soroti road and a number of local roads to the district. There are a number of key taxi and bus pick-up and drop-off points located on the edge of the south-west of the site to serve the centre.

Figure 4.1 Kachumbala Health Centre 3 in January 2014

EFOD SWW decided to assess the improvements needed, and Dan Flower, who had operated as EFOD's architect for some years, sought out the local data necessary. Copies of the Ugandan Department of Health standards for maternity units were obtained from the district engineer, and discussions held with the chair of the local health committee and the staff, together with the head of the council and the regional district commissioner while the team was still in the area. The need for humanitarian aid was clear, and EFOD accepted the challenge to make a difference for the maternity provision for the women of Kachumbala.

Research

A brief to design and construct a new maternity unit as an extension to the medical centre was developed, to improve local maternity capacity and increase the safety of both the mother and child.

The key objectives were to

- reduce the number of home births
- increase medical supervision and reduce referrals to local hospitals
- ensure new mothers remain in the centre for 24 h, in line with World Health Organization (WHO) guidelines
- promote the importance of primary healthcare.

The original medical centre was a single-storey structure built from clay-fired bricks with a timber truss roof clad with corrugated sheets. A mains water supply was available, although unreliable, supplemented by a borehole. Pit latrines were available at the rear of the site. Initially there was no mains power, and small solar panels had been installed to charge batteries for lighting and small electrical appliances.

With three births per day and only three recovery beds in the old unit, mothers were often on the floor or discharged soon after birth, contravening Ugandan health guidelines. The alternative was a 20 km trip to the nearest regional hospital – a journey many were reluctant to make due to lack of funds or transport.

EFOD's objective was to provide a simple to build, use and operate unit, increasing privacy and dignity for patients, promoting cleanliness and improving working conditions; a bespoke solution allowing the building to respond to user needs, site constraints and its environment.

The existing centre is set on a large site, but construction of the new extension with minimum disturbance to the operation of the facility was clearly necessary.

Scheme design

EFOD invited Carole Bell and Julie Jenkins, senior midwives responsible for maternity provision in the Hywel Dda Health Authority in West Wales, to join the team and assist with the design. As the scheme developed, contact was made with Dr Kathy Bourgoine, who specialises in premature births, and her husband Dr Adam Hewitt-Smith, an anaesthetist, working as long-term volunteers in nearby Mbale Regional Referral Hospital (Mbale RRH), and run the Born on the Edge charity. All provided excellent advice on the layout of the unit, maternal care and equipment availability in Uganda.

EFOD had been building structures in the area for a number of years and so was familiar with local building practices.

Layout

The design was developed during 2014 in consultation with health leaders in Kachumbala and the UK, with new layouts implementing best practice in the separation of patients and waste flow, and providing dedicated spaces for pharmacy, drugs storage, administration and cleaning facilities.

The unit was designed to maximise passive ventilation, which is key to the comfort of mothers, babies and staff. EFOD had designed and built a medical centre in Soroti, 100 km away, 4 years earlier, with a mono-pitch roof and significant high-level ventilation. It proved partially successful, but airflow through the building has a tendency to deposit dust within medical rooms. So, the design for Kachumbala was modified to introduce central delivery suites and infrastructure, with corridors either side. Terracotta block external walls permit air movement, resulting in the free flow of air, drawn from delivery suites and service rooms through ceiling vents, with dust settling in the corridors. This design overcomes the need for mechanical plant to move air.

This layout also provides shading against direct sunlight in the delivery suites.

Two delivery suites were included, to allow for up to six deliveries per day, sharing a sluice room and pharmacy. One corridor provides access for patients and staff, the other for waste disposal and access to staff toilets. A ward to accommodate eight beds was necessary for the centre to meet WHO guidelines for duration of stay. An isolation room with two beds was added for patients suffering from malaria.

An L-shaped single-storey structure with mono-pitch roofs was chosen, sitting adjacent to and behind the existing provision, to minimise disruption during construction, and to provide a reception area for the new unit. Figure 4.2 shows the layout of the new unit.

Material availability

Burned clay bricks are in extensive use in Uganda. Clay is readily available, and this can be pressed into moulds and fired, providing income for brick makers. However, 70 t of timber is needed to fire a substantial (10 000) brick pile for 5 days, resulting in the loss of trees and generating a significant volume of carbon. The result is a relatively low-grade brick, usually of variable size because of the home-made moulds in use, and the need for wide mortar joints when laid. EFOD started to use sustainable interlocking stabilised soil blocks (ISSBs) in 2009, and has promoted their use on all subsequent schemes in Uganda. Murrum, a natural mix of sand and stone, underlies the site, providing a competent founding layer for the building. The material excavated from the deep septic tank and soakaway pits was riddled, mixed with cement in a controlled manner, and pressed into blocks in a Ugandan-designed block press before drying in the sun.

Locally manufactured concrete hollow blocks were available and used to form reinforced concrete columns. Materials were purchased in the village of Kachumbala, 2 km away, when available, and from Mbale, 20 km away, for specialist materials such as terracotta blocks, or when large orders were required.

Figure 4.2 Layout of the new maternity unit

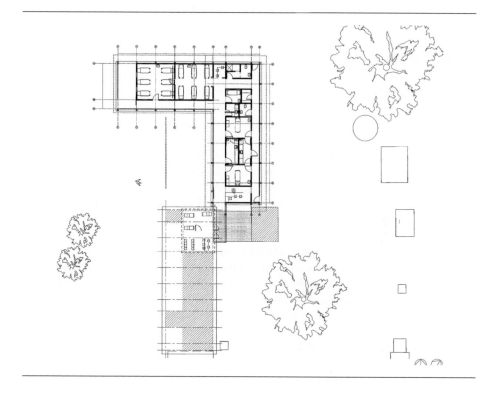

Funding

EFOD estimated a construction cost of £110 000, to cover materials, labour, and travel and subsistence for UK volunteers to supervise the work. This is a substantial sum for a small charity to raise, and required a number of initiatives over 4 years by all involved. Since the challenge had been issued to the First Minister for Wales, the Welsh Government was approached for support, and generously provided £20 000 under its Wales for Africa initiative to assist the training of apprentices and engineers, funding the cost of their flights.

This was a good start to the scheme, and once the mill and grain store in Kachumbala town built by the same team was completed, construction of the maternity unit started in March 2015, with a formal ground-breaking ceremony by the senior local counsellor, William Wilberforce. Photographs of early construction work were then used to assist with further fundraising initiatives, which included formal dinners, sponsored cycling events, and grant applications to professional bodies. Cyfle in South West Wales arranged annual 'village day' events in Ammanford, apprentices undertook individual events at their local Coleg Sir Gâr campus, and companies seconding staff to travel to the site were persuaded to provide support. Funding dictated the rate of construction, resulting in four phases over a period of 32 months.

Dan Flower's employer HKS, a US-based architectural practice with a London office, accepted the scheme as its first major corporate social responsibility project, through its Citizen HKS initiative. It donated 2000 h of professional time to assist with the architectural design, and then had a major fundraising initiative among its 1000 staff worldwide, resulting in a \$45 000 grant that completed the fundraising.

Construction

Upon completion of the design, and approval granted by the Ugandan authorities, EFOD engineers, technicians and qualified apprentices from south Wales travelled out to Uganda in pairs for a period of just 2 weeks to manage the construction. The foundations were excavated by hand and cast, together with the ground floor slab, over a period of 4 weeks in March 2015 by a local labour force of 40. The UK supervisors were responsible for planning the work, purchasing materials and managing the finances. This provided excellent experience for all. The EFOD volunteers and the local workforce shared their skills and learned from each other. The workers were paid a fair wage at the end of each day, and were fed from an on-site kitchen. Little mechanical plant is available locally, and so none was used since hand work increases employment.

Good building and health and safety techniques were taught through toolbox talks. Personal protective equipment was issued to all workers. Coloured T shirts proved popular with all, and helped to identify genuine workers for payment at the end of each day. Scaffolding and ladders are not available, so timber-framed platforms were made on site with safety in mind. A lower ladder extension was developed, to mimic an adjustable ladder.

Phase 2 of the scheme was undertaken in autumn 2015.

A team of three was employed to make ISSBs from the excavation arisings of the septic tank and soakaway pits, forming up to 500 per day, and laying them out to cure for 7 days before use.

Reinforced concrete columns were formed using infill to voids in locally cast hollow concrete blocks. Walls were constructed using ISSBs, and the reinforced concrete ring beams were cast. Figure 4.3 shows the unit under construction.

Once further funds had been raised, the roof was added in 2016 during phase 3. Timber trusses were assembled on site using locally purchased timber, with the roof truss spacing set to readily available purlin lengths to reduce wastage. The EFOD volunteers realised that materials purchase in Africa differs significantly from the UK. Rather than an email to a supplier with a pitch and a span, to be delivered to the site complete, trusses were designed by Cardiff-based engineers, and timber bought from timber suppliers in Mbale. Each piece had to be selected by hand, and loaded into a locally hired vehicle. Loading and unloading were considered an additional activity by the labour force, for which additional payment was expected. Only then could assembly begin, with the local carpenters demonstrating their admirable skills.

Walls were rendered, and external terracotta panels erected. In addition, ceilings to the delivery suites, offices and wards were added, and plumbing and the first internal fix of the electrics were also undertaken.

Figure 4.3 Construction of the maternity unit

For the first time in Uganda, EFOD witnessed a scaffold decking for ceiling work, using recycled shutters from the ring beams.

Phase 4 to complete the scheme started in mid-2017. High-quality hardwood windows and doors were fabricated in Mbale, although there was a lesson to be learned. Negotiating a very competitive price resulted in slow manufacture, and seriously affected the rate of delivery, and so delayed the programme.

Carpenter Owain Phillips had worked on the scheme during an earlier phase, and was changing employment. He volunteered to spend a further 6 weeks on site, which was a significant benefit during the final phase of the works. Local tilers were employed to lay the floors, painters to decorate, electricians to finish the wiring and plumbers to complete the plumbing. Continuity of supervision at this stage was crucial to obtaining a high standard of finish.

Energy

At the project inception in 2014 there was no mains electricity to the existing health centre building, which meant that torches or mobile phones were often necessary for night-time deliveries. A mains connection was added as the build progressed through 2016–2017, although this was still intermittent. The necessity for an alternative source was demonstrated in the week of opening the maternity unit in November 2017, when the mains power was off for a full week.

Because of the scarcity and unpredictable nature of resources in the region, the unit is designed to operate as a net-zero building, to maximise its ability to serve the local community. Ring mains and lighting circuits connected to the mains were installed throughout the unit for lighting, and to power small appliances and charge mobile phones. A separate DC system powered by solar panels charges batteries, providing light to critical rooms 24 h/day and running a specialist 12 V vaccine refrigerator. The AC and DC circuits are independent, since experience from a previous project demonstrated that careless switching between the two could lead to serious damage to the system. There are no turning fans or air systems in the building, and passive ventilation is regulated by operable louvre windows and ceiling grills in each room, allowing ventilation and cross-flows of air.

Water

Two rainwater collection systems were installed, one local to each mono-pitch roof. A 5000 l external storage tank with an electrical pump feeds a 2000 l header tank, to service flushing toilets, a shower and wash hand basins. An additional 3000 l collection tank is for community use. The mains water to the centre is used to supplement supplies during the long dry seasons. Waste water is routed to a new septic tank, with a soakaway system at the edge of site, in the absence of a local sewage network.

Since the new unit has been operational, figures indicate a 48% drop in electricity consumption and an 18.5% drop in water consumption. Although the centre has doubled in size, this significant reduction is attributed to the improvements implemented.

Equipping

Medical equipment such as delivery and ward beds were obtained from Joint Medical Supplies in Kampala. Savings were achieved with joint ordering with the Mbale RRH through Dr Adam Hewitt-Smith. Furniture such as bedside cabinets, desks and storage cupboards were made by the carpenters at Mbale hospital, providing cost savings over imported items, and work for local carpenters.

Opening

The unit (Figure 4.4) was formally opened by the senior local counsellor on 2 November 2017, with a party for almost 2000 people, with music, plays and prayers. The first birth was in the morning of 3 November, with the safe delivery of Simon to mother Jessica.

The unit is a fine and well-equipped building (Figures 4.5 and 4.6). Passive ventilation ensures comfort for the occupants, and two delivery suites ensure expectant mothers can deliver in relative comfort. The ward is of sufficient size to accommodate mothers and newborn babies for up to 3 days. The old maternity rooms have been converted into a teaching room and a family planning room.

The unit was completed in just under 4 years from challenge to opening, within the original £110 000 budget.

Figure 4.4 External view of the maternity unit

Figure 4.5 The delivery suite

Figure 4.6 The ward

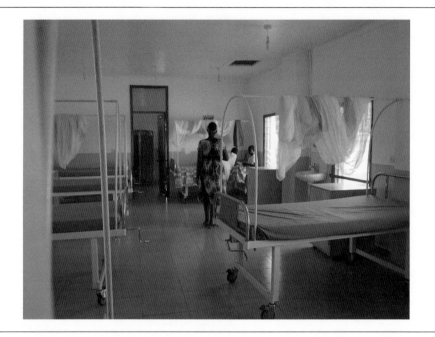

Benefits

The women of Kachumbala now have access to a maternity unit fit for a new mother.

A labour force of 40 was employed for a total of eight working months during four phases of development. All workers were paid a fair daily wage, and provided with breakfast and lunch prepared in an on-site kitchen by local cooks. The workers learned new construction techniques and received training in site safety from UK-based engineers and skilled craftsmen, and shared their building skills in return.

The scheme funded five midwives in the unit to receive additional training in the neonatal care of premature babies from Dr Kathy Bourgoine in nearby Mbale RRH. This has increased the confidence of mothers in the area, encouraging them to give birth in the unit. Midwife Julien Ikaraat said she has not lost a child during a delivery since completing this course. The project partners have funded additional training for midwives to increase their skills, aiming to reduce the number of referrals. Problems and delays associated with transfers to hospital have caused the occasional death of mothers and babies. Perhaps because of its growing reputation in the area, the medical centre has recently received an ambulance to assist with the transfer of serious cases – a life-saving addition.

Twenty young UK construction professionals and skilled workers made a total of 34 visits to share their skills on site, and while there developed new management and supervisory techniques. Tom Hawken used the experiences gained as the scheme's project manager to assist his successful application to become a Chartered Civil Engineer.

Two qualified apprentices realised they were capable of taking much more responsibility following visits, and have gone on to obtain degrees in construction and management. Leyton Davies had planned to continue as a bricklayer; he is now a building inspector. Owain Phillips is now a project manager. Others have benefited from the varied experience on offer. Joanna Capela, an environmentalist, spent 3 months on site managing the final finishes and installation of medical equipment, learning much about finishing and equipping a medical facility.

Shaun Edwards, an electrician, joined Cyfle Building Skills on a Thursday, flew to Uganda 4 days later in support of EFOD, and was on site on the Wednesday morning to wire the facility, when it became apparent that a failure of the backup power supply in the neonatal unit at Mbale RRH was leading to two child deaths each time there was a power cut. He was able to visit the hospital, test the system and repair a faulty inverter unit, to ensure the backup system worked again, preventing further deaths, before returning to Kachumbala to wire the building.

All involved in the design, construction, equipping and opening of the Kachumbala Health Centre 3 Maternity Unit attest to the benefits gained from their involvement in the scheme.

Monitoring

EFOD and HKS have been carefully monitoring the performance of the unit since opening to see if the new facility is meeting its objective for the safe delivery of babies in a well-equipped unit.

One of the authors visits Uganda two or three times a year to evaluate existing and new schemes, and has been able to maintain contact with medical centre staff in the unit, discussing progress with the maternity staff and local health board leaders. Weekly reports have been obtained for a period of 24 months following opening, and the results have been analysed by the HKS research team.

The number of deliveries per month has increased by over 40%, reducing home births in the area.

The number of mothers referred to local hospitals has reduced by over 18%. The duration of stay in the unit post-delivery has increased to between 24 and 72 h. Infant mortality has reduced to just one over 24 months.

The unit is well run, and has become a major midwife training centre: 24 trainees were in residence for a week in January 2020 to gain work experience. The number of midwives has increased to five. The satisfaction rating of patients is 9.77 out of 10. (The installation of a TV in the ward was necessary to increase the score!)

The isolation ward has been converted into a neonatal unit for pre-term babies.

Akello, a new mother, said, 'My heart just likes this place because it is really clean and I feel very safe.' And according to a member of staff, 'Some mothers say they do not want to be discharged, and others wish to have more babies so they can deliver here'.

Conclusions

All the key objectives have been met, reducing the number of home births, increasing medical supervision, reducing referrals to local hospitals, ensuring new mothers can remain in the centre for 24 hours (in line with WHO guidelines) and promoting the importance of primary healthcare.

So much more can still be done. Following the opening of the new maternity unit, the local authorities plan to upgrade the status of Kachumbala's Health Centre from HC3 to HC4. The centre is currently run by a clinical officer serving a population of 55 000. The upgrade will result in the provision of a full-time medical officer (a fully qualified doctor) for the community, but not before an operating theatre is built on the site.

There is land, there is the will, and EFOD SWW is considering further involvement.

Acknowledgements

The authors are grateful to all those involved, both named and unnamed, for their commitment delivering this wonderful project, including

- engineers, technicians, and skilled craftsmen and women from Teso, Uganda, and from Wales
- funding from TRJ Builders, TAD Builders, BPEC, CITB Wales, Dawnus, SFS, the Welsh Government, Citizen HKS and HKS staff
- the medical staff and midwives working in Kachumbala's medical centre
- Dr Kathy Bourgoine and Dr Adam Hewitt-Smith, who run the charity Born on the Edge in Mbale saving the lives pre-term babies.

ENVIRONMENTAL IMPACT MITIGATION: ADAPTING SHELTER AND SETTLEMENT SOLUTIONS IN BURUNDI

Antonella Vitale and Lewis Kelly

Introduction

Construction has been a key driver for the local exploitation of natural forest resources in developing contexts, for the sourcing of materials such as timber and bamboo, the harvesting of energy, water and fuel wood for production processes, and the exploitation of earth, sand and clay resources from marshland habitats (Alam and Starr, 2009; Fuwape, 2003; Skiadaresis *et al.*, 2019). Driven by similar resource needs, large-scale shelter programmes can have acute negative consequences on surrounding forested regions (Ashmore and Fowler, 2009; Zarins, 2018), and it has been seen that degradation of local forest areas increases in the proximity of refugee camps (Win Thin, 2018).

While many issues behind this degradation are due to socio-political conditions and resource management on a local level, the shelter sector is beginning to respond to its inherent responsibility for the impact of construction from a more globally informed perspective.

Addressing the environmental impact of materials such as cement, steel and plastics, natural materials such as timber are widely promoted as a sustainable construction choice in architecture and construction, due to their lower embodied energy and carbon emissions from production processes and potential climate benefits of carbon sequestration during growth (Matard *et al.*, 2019; Peñaloza *et al.*, 2016). This is complemented by an increasing appreciation of the sustainability benefits of procuring materials locally, reducing emissions and energy from transportation (Escamilla and Harbert, 2015). Addressing climate goals and the plastic waste crisis, locally procured organic materials as well as earth-based techniques have hence been promoted by shelter actors and academics within environmentally appropriate shelter design, although sometimes without a thorough analysis of the impacts on the local environment (Kuittinen and Winter, 2015; Matard *et al.*, 2019).

This is a problematic conflict between global and local definitions of environmentally appropriate construction, and an increased vulnerability of settlements to natural hazards is being seen worldwide as a consequence of local deforestation and resource extraction. The correlation between deforestation and the increased severity of natural hazard events such as floods and landslides has been documented across different contexts (Bradshaw *et al.*, 2007), as have been the devastating effects on settlements in both building damage and loss of life. The benefits of forests for climatic regulation, landscape stabilisation, soil protection and water system regulation have long been appreciated, and their resilience benefits are being increasingly understood – including water retention and decreased water run-off (EEA, 2015), reducing the possibility and severity of floods and mitigation of drought patterns through regulating the run-off seasonally (Sfeir-Younis, 1986), and the stabilisation of earth, hillsides and riverbanks (Stokes *et al.*, 2008). Even conversion of forest to agriculture is seen to cause a severalfold increase in soil erosion, leading to increased levels of landslides. In light of the vulnerability that environmental degradation can bring to settlements, it is paramount to stress the need for greater contextual insight for informed material choices, to ensure that shelter and settlement practices do not exacerbate the vulnerabilities they seek to protect people from. Unfortunately, it is not rare for big emergency responses, where shelters need to be provided at scale, to exacerbate negative impacts on an already fragile environment.

It is worth recalling the lessons learnt during the tsunami response in Aceh, Indonesia, in 2004, where material sourcing, design optimisation and construction techniques were not considered crucial by most implementing agencies. Most used locally procured materials without realising the environmental implications of their decision, particularly when replicated at scale. The use of timber, as construction material, was initially taken to be a culturally accepted and default option, only to later find the impossibility of verifying its legal sourcing, requiring a drastic shift to non-locally purchased timber (Da Silva, 2010). Another lesson comes from the earthquake response in Haiti in 2010, where construction materials needed to be almost entirely imported, and where the very scarce wood available was sourced as fuel. The earthquake produced 40 million cubic metres of debris, which was removed and stored randomly, resulting in further hazard, pollution, infrastructure congestion, and delayed distribution of aid (UNEP, 2010). Coordinated removal, recycling and reuse of debris would have contributed to the necessary stabilisation of land, as well as its raising in critical coastal and riverbank areas.

This sub-chapter is based on a study undertaken by the International Federation of Red Cross and Red Crescent Societies Shelter Research Unit (IFRC SRU) in 2019 in collaboration with CRATerre and the Luxembourg Red Cross, which surveyed the local building practices used for shelter programmes in Burundi, assessing their impact on the local natural environment as well as drawing insights into the influence of socio-economic and political factors behind their use.

Burundi is one of the poorest countries in the world, with an average annual income of $215 per inhabitant since the 1990s (World Bank *et al.*, 2017). This East African country offers insights into a situation in which a culmination of socio-economic and political factors has resulted in a degraded natural landscape vulnerable to natural hazards and climatic threats. Burundi is still recovering from extended episodes of violence in its recent past, including the civil war, which lasted from 1993 until 2005 and left 300 000 dead, 400 000 refugees and 880 000 internally displaced people. The country has seen high economic instability and inflation in recent decades, raising the price of construction materials over short periods of time. Despite high poverty levels, the construction costs in land-locked African countries such as Burundi are higher than most other countries in the continent, due to the price of transportation from abroad. Compounded by the volatility in the price of petrol and the frequency of stock breakages in transportation, the lack of reliability of these expensive imports has resulted in a high dependency of Burundi's population on its own natural resources, agriculture and building materials (World Bank *et al.*, 2017). This dependency on internal resources continues to be an important trend, and hence their responsible management is especially important for Burundi´s development in the coming years.

Forests, deforestation and resilience

Despite the UN Sustainable Development Goal 15, 'Life on land', and the UN 2030 Global Forest Goal 1, to protect and restore forests and increase their area by 3% by 2030 (UN DESA, 2019), local deforestation continues to be seen worldwide, and Burundi has been a key example of the link between an unstable political landscape and devastation of its natural habitats. With little environmental safeguarding in place, its economic growth during recent periods of conflict and instability has been largely based on the depletion of its natural capital, with its forest cover reducing from an estimated 30–50% to just 6.6% today (World Bank *et al.*, 2017). This loss has included tropical forests from logging as well as marshland habitats from the extraction of clay. Burundi's deforestation rate has been three times higher than the Sub-Saharan African average over recent decades, peaking during the civil war, when the country saw the highest deforestation rate in the world of about 9% (AFF, 2011).

Burundi's forests have provided essential ecosystem services, contributing to the resilience of the main urban areas of the country, including regulating the water cycle, protecting watersheds, preventing soil erosion and maintaining the integrity of soils, and stabilising hill slopes such as those upstream from the capital city Bujumbura (World Bank *et al.*, 2017). Deforestation has compromised the resilience of these landscapes, resulting in more severe and frequent flooding and landslides, damaging houses and infrastructure, and claiming lives and livelihoods, exacerbated by the more intense climatic events seen in Burundi in recent years. The country is regarded as being among the least prepared globally for climate change (World Bank, 2018).

Forestry management and demand for timber and fuelwood

The principal drivers of deforestation in Burundi are the demand for timber for construction, and the harvesting of wood for fuel in producing fired bricks, cooking and heating. Some 90% of the population is reliant on wood for cooking and heating (ProAct Network, 2009). With increased housing needs from the rising population, communal and state forests have been logged to meet the demand for timber for residential construction, at times with the complicity of local authorities (World Bank *et al.*, 2017). The deforestation has also been partly due to the high dependency of the population on internally cultivated agriculture, and hence a high demand for agricultural land, combined with poor land management, which has further reduced soil quality, lowering agricultural productivity and increasing the demand (World Bank *et al.*, 2017).

Due to political turmoil and insufficient resources, there has been a lack of responsible forest management in Burundi to regulate environmental degradation. The Burundi Government made many efforts to undertake tree plantation programmes over past decades, particularly in the 1970s and 1980s, with support from funders including the EU, World Bank and United Nations Development Programme (UNDP) among others (AFF, 2011). However, many of these efforts were lost through the war from 1993, with heightened exploitation and forest burning during this time. In acknowledgement of the environmental issues facing the country, the Burundi Government has since implemented environmental legislation to regulate deforestation activities and encourage the responsible management of forests, although it has been noted by ProAct Network (2009) that this has been largely ignored and unenforced due to insufficient human and financial resources, and the last national forest inventory was undertaken in 1976 (AFF, 2011).

While previous afforestation efforts by agencies encouraged mixed-species habitats, agroforestry plantations in the country today are seen to favour monocultures of exotic species such as *Eucalyptus*, which are able to supply the demand for timber quickly due to the species' fast growth rate (ProAct Network, 2009). This can be problematic due to the high quantities of water needed by *Eucalyptus* and its effect on depleting nutrients from the soil, making it difficult for other species to survive nearby (Tererai *et al.*, 2013). A reduction in biodiversity from plantation monocultures restricts the ability of other plants to contribute to the preservation of soil integrity and robustness of the landscape. However, their benefit of high biomass is appreciated, and while some politicians recommend the uprooting of this species to replant endemic trees, it was discussed at a regional workshop in 2010 in Bujumbura that these species have value and should be more carefully planted in rocky terrains that would not endanger the surrounding landscape (AFF, 2011).

Local versus sustainable materials and construction practices

While on a general level, with respect to global warming potential, the promotion of locally procured natural materials is well founded, it can be seen how problematic this generalisation could be in contexts of extreme natural degradation, poor forestry management and levels of poverty, which exert growing pressures on weakened landscapes. It therefore seems critical that these valuations of environmental impact and contribution to landscape vulnerability become key considerations in the selection of materials for construction, both locally and within shelter programmes.

In response to the periods of conflict in Burundi, numerous shelter and settlement programmes have been established by different organisations to house the displaced populations. These have encouraged construction using the local building materials found in the country, supporting self-construction by beneficiaries according to local techniques.

The IFRC SRU investigated the different approaches to construction found in the country, to understand their impacts on the local environment and help inform more responsible material choices in shelter programmes. The study focused on the provinces of Kigwena and Muyinga, with 174 and 1200 shelter programmes, respectively.

The survey revealed that significant amounts of timber were used for the roof structures of houses, as well as for timber frame houses, outbuildings and extensions clad in wooden planks. This was the case both locally and in settlement programmes. The shelter programmes themselves employed a mixture of these techniques, but also included compressed-earth blocks (CEBs), a masonry technique not seen in local construction practice.

The roof structures and connections were often inefficiently made, consuming much more material than necessary and highlighting the need for capacity building on improved ways of building more solid roof structures with less material. The use of fired roof tiles was found to require steeper roof pitches than corrugated galvanised iron (CGI) sheets, requiring larger timber structures and greater reinforcement to cope with the additional weight of tiles compared with the metal sheets. While it was observed that efficiency savings could also be made in the use of timber, there was also a clear correlation between the affordability of roofing and the indirect demands on the quantity of timber to support it.

Understanding the impacts of the use of timber goes beyond the quantities involved, to include the sourcing and species of the material. An impact study was undertaken by ProAct Network to evaluate the environmental impact of shelter programmes in Burundi by a key non-governmental organisation (NGO) at that time, whose operations included 13 000 shelters and 800 classrooms for both returnees and the local population. The NGO used a great deal of timber in construction, with around 108 trees cut for the construction of a temporary classroom, although the study praised the organisation for its consideration of environmental issues in shelter policies, which included ensuring that construction wood sourcing was tracked. The study did, however, note a potential issue regarding the impact of the species sourced itself: given that indigenous species were considered inappropriate, the large quantities of wood for the camps were procured from *Eucalyptus* plantations. Sourcing from such plantations may seem like the most sustainable option, given that it grows quickly and can be replenished. However, this species, when irresponsibly managed, has the potential to exacerbate environmental issues due to its high water demand, impact on soil quality and the lack of biodiversity present in its monocultures. The study additionally noted that the NGO at that time did not engage with any afforestation programmes to offset this timber use (ProAct Network, 2009), which could be seen to conflict with a long-term approach to the resilience of local communities and settlements to natural hazard events.

Adobe and wattle and daub construction

In the early 20th century, indigenous construction practices in Burundi of reed and thatch huts evolved to incorporate techniques using earth construction. Adobe construction using earthen blocks emerged at the end of the 19th century, which gained widespread popularity in the 1950s and spread through the country, similarly to the construction of walls using earth held together with latticed bamboo and timber structures, known as wattle and daub. These construction techniques were found to be very prevalent today across the territories surveyed. They are affordable and technically accessible, and largely produced by local people without technical support.

On an environmental level, the adobe bricks used locally with little cement can have a negligible environmental impact and very low carbon dioxide emissions, particularly if the earth used is from the site itself, with no transportation involved. They also have a minimal impact on forest resources, using timber only in the form of reusable wooden moulds.

Adobe brick construction does, however, have structural weaknesses, and so there are limitations on its recommendation for shelters, particularly in a seismically active country such as Burundi. Unstabilised adobe construction is prone to collapse, as well as to erosion over time from rain and flooding. There is a restriction on certain earth types, such as the black earth around Lake Tanganyika, which contains high levels of organic matter, making it too porous for use in adobes. Earth with excessive clay content also carries the risk of cracking during the drying process, and so a careful assessment of the local soil type must be done prior to the choice of this construction method, to ensure durability. This complexity is compounded by the need for continual maintenance and high protection from water penetration. The structural weakness of the material has led to the prohibition in Burundi law of adobe blocks for use in larger permanent public buildings (ProAct Network, 2009), with promotion of the use of other sturdier techniques such as fired bricks in local construction.

Wattle and daub construction has a similarly low carbon dioxide emission to adobe, with resources extracted locally and little need for cement, due to the way that the timber weave binds the infill material together. Wattle and daub does, however, use large quantities of timber to give the infill material stability, with one system of substantial timber poles used to support a secondary system of woven branches every 10 cm vertically. This quantity of wood would imply a considerable reliance on harvesting trees if employed in large-scale settlement responses, either intentionally or through encouraging self-construction by beneficiaries using this technique.

Fired bricks and tiles – an environmental menace

Fired artisanal bricks and roof tiles were introduced to Burundi in the early decades of the 20th century, and are now popular and widespread throughout the country, as a low-cost and accessible construction technique that can provide good stability properties if manufactured well. Local people are provided with clay from deposits exploited by the authorities in return for an annual tax, so they are produced in huge quantities in Burundi, using traditional firing techniques. While this masonry technique does not directly consume organic forest resources, it is important to understand the indirect environmental impacts these materials have through their sourcing and manufacture.

Due to the widespread use of wood to fuel the firing of these materials, fired bricks and tiles have had a highly detrimental impact on forest reserves in Burundi, thought to be around two-thirds of the annual loss (DDC, 2016). They demand enormous amounts of wood to be burnt for their firing, estimated to be around 1.3–2.9 million tonnes/year (Savary, 2011), so much that the Burundi Government has declared their use to be of great concern on a national environmental level and called for the use of alternative techniques (French.china.org.cn, 2018). This consumption rate is partly due to the inefficient kilns used across the country, and their carbon dioxide emissions were found to be around 12–15 times that of adobe blocks. Requiring high clay content, fired bricks and tiles have also have a key role in the unsustainable exploitation of vulnerable marshland habitats, with no restrictions imposed to limit this degradation.

While the government has been promoting the abandonment of this construction technique, fired bricks are still used widely for house construction, both in local practice and in shelter and settlement programmes. These fired clay materials are affordable to local people, a key reason behind their proliferation over other less impactful materials. While there has been a promotion of CGI metal roofing across the settlements surveyed, fired roof tiles are used widely in construction across the regions, primarily in Muyinga province, principally due to the affordability of this material in comparison with CGI metal roofing. This is an important consideration, given the economic situation of many vulnerable groups in Burundi, and it has been noted that due to inflation the price of CGI sheets has risen 60% in 4 years, and the nails necessary for installation by 70%. These economic reasons have a large influence on the local choice of fired tiles over CGI roofing – important when promoting construction by local people themselves, as was done in Muyinga.

With 1.5–3 m^3 of wood required for 1000 traditional bricks, and at least a 25% breakage rate, it is estimated that a potential 60–90% of the energy needed to fire bricks and tiles could be saved through improvements to stoves. There are some programmes in Burundi working to provide stoves with improved efficiency and to refurbish existing cooking stoves across the country, while making improvements to allow the use of renewable biomass from agriculture as opposed to wood fuel. These include Burundi Quality Stoves (UNFCCC, 2014) and the PROECCO programme, both of which are working to provide more efficient ways of firing bricks (Skat Consulting, 2017). While offering improvements, these programmes are not scaled-up across the country for widespread use, and an important capital cost would be involved in doing so. The use of biomass from agriculture is also not necessarily an answer to the use of wood fuel, as it has been widely noted that agriculture is a key driver for deforestation, although making use of the waste by-products of this process would definitely offer theoretical potential for reduced depletion of forested resources for fuel in the production of materials.

Compressed earth blocks

To address the environmental threat presented by the intensive production of artisanal-fired bricks, several alternative materials were introduced to Burundi in the 1980s, such as CEBs. CEBs do not require firing, and hence have little impact on forest reserves. They are stronger and more durable than fired bricks, can be made in situ, and so do not require transportation, and their environmental impact is lower in terms of embodied energy and emissions over their longer life cycle. While carbon dioxide emissions are higher for CEBs than for adobe blocks given the cement used, variations of this technology have included interlocking CEBs (Eires *et al.* 2012), which also eliminates the need for mortar/cement and its associated embodied energy and emissions. Due to this potential, CEBs are seen across the provinces in shelter programmes in Burundi, as well as in sustainable architecture projects such as the Library of Muyinga by BC Architects and Studies (Grozdanic, 2014). They are promoted by the Burundi Government as a less environmentally damaging masonry technique, with the settlement programme in Muyinga requested by the government to use this material for the construction of all shelters. These blocks do, however, require a large amount of water for curing: 8 l per day per block for 14 days, which results in 112 l per block. To put this in context, identical shelters built from adobe or fired bricks would need only 600 l per 1000 bricks, or 0.6 l per brick, around one 1/185th of the volume for a CEB. This is not a negligible quantity in contexts that experience drought such as Burundi. Water consumption could be improved, but until then a need remains for more studies to be undertaken to ascertain the extent of the impacts that this technology would have on water resources if rolled out across the country.

Despite the promise that CEBs offer for some lessening of the impact on forest resources, there has not, however, been an uptake on a large scale of CEBs. This could be partially due to the price: the IFRC SRU study found that CEBs cost around twice as much as both adobe and fired bricks. The lack of uptake locally may also have to do with other complexities, such as the need to maintain the hydraulic presses, the demanding labour requirements, or the fact that the presses are owned and operated by an association, rather than facilitating widespread manufacture by the inhabitants themselves. In the settlements surveyed there was a far greater proportion of CEBs than in the country as a whole. However, the much higher cost of this material was not budgeted for originally, resulting in a large number of shelters not being completed.

Resilient shelter and settlements – conclusions on responsible material choices

The sustainability of forest resources cannot be generalised, nor can the local procurement of construction materials to mitigate transportation impacts on the environment. While locally procured timber use may have global appreciation for low embodied energy and embodied carbon, the impacts on the local environment in countries such as Burundi must be considered. Embodied energy and embodied carbon are vitally important measurements of sustainability in construction, but cannot be isolated from other parameters and variables. The use of local natural materials can lead to local deforestation and landscape vulnerability in countries such as Burundi, where widespread degradation is threatening the resilience of populations. An understanding of the effects on local resources may mean that there are trade-offs to be made until programmes are integrated with responsible environmental management.

The use of fast-growing plantation species may seem like a solution, but this cannot be generalised as a more sustainable alternative to the use of endemic tree species. Feeding plantation tree markets can encourage the planting of exotic-species monocultures, which deplete soil of its integrity and nutrients, destabilising landscapes.

In avoiding timber, earth construction is not necessarily free from impact on forest resources. Materials demanding clay content are responsible for the degradation of other natural habitats such as swamp areas, and the burning of the clay to make bricks and tiles is one of the key drivers of wood consumption. The choice of alternative construction must be made with an understanding of the sourcing and manufacturing processes, which goes beyond an understanding of the material itself.

Pursuing 'sustainable' technologies must be done with a consideration of secondary environmental burdens; although certain materials such as CEBs have little impact on forest resources, their high water demands could put great pressure on drought-prone countries such as Burundi.

Certain materials that are regarded as less sustainable due to their production processes and the need for transportation (e.g. CGI sheets) can, in fact, have a lesser direct impact on the local environment than their alternative – fired tiles in the case of Burundi. CGI sheets also allow for a reduction in roofing structures compared with tiles, therefore conserving timber. Materials viewed as less sustainable globally, or those imported from elsewhere, may be a worthwhile trade-off while local landscapes are regenerated.

Sourcing bricks from environmentally oriented material innovation programmes, which encourage the use of higher-efficiency stoves for firing bricks and tiles, could help to reduce the local environmental impact on forests in Burundi. This may mean greater transportation distances to the source compared with established facilities and increased greater carbon dioxide emissions, but this trade-off against environmental impact may lessen over time as the market for such bricks is bolstered and supported to expand the use of improved stoves more extensively across the country.

The economic affordability of sustainable materials is clearly a vital factor in shifting local construction practices away from more cost-effective but environmentally damaging techniques, and will remain a limitation until this is addressed. When CEBs were used in shelter programmes, no effect was recorded on local construction practices, as CEBs remained unaffordable for most people. Political instability too can mean that materials cannot be easily procured internationally, which might require promoting internally sourced materials for the resilience of local communities.

The additional costs of sustainability have been recognised, whether in manufacturing processes, the establishing of improved material production facilities, or in investing in environmental regeneration efforts. While informed decisions can be made about environmentally appropriate material choices, often funding, as well as economic interests, dictate the choices made. Shelter programmes should have funding earmarked for environmental aims, to allow for the support of more appropriate responses despite any immediate and apparent additional cost, which may form barriers in standard practice. While the local environment must be acknowledged as a vital part of ensuring long-term resilience to communities, funding needs to reflect the additional allowances required to nurture a more resilient landscape of shelter provision for more resilient communities.

To address such limitations and offset any unavoidable local impacts of construction material choices, there is the potential for shelter programmes in countries such as Burundi to more actively engage with ongoing environmental schemes. This could involve ensuring that timber is sourced from responsibly managed plantations, appointing their environmental experts from within shelter programmes to assess the location and species of trees, to ensure that they have not been planted at the expense of other species and the integrity of landscapes. Some large-scale programmes exist in Burundi, such as the Central Africa Forest Ecosystems Conservation and Environmental Monitoring and Policy Support projects from the US Agency for International Development, partnership with whom the shelter sector could benefit from in order to draw specialist knowledge about the resilience benefits of local landscapes, and to make informed decisions about material sourcing. This could help mitigate deforestation-induced vulnerabilities of landscapes and settlements, as well as vulnerabilities of individuals dwelling within them. There exist numerous local programmes within Burundi seeking to conserve and regenerate forests, and to address climate change resilience through better management of natural resources, such as the Association Burundaise pour la Protection de la Nature, and the Organisation de Défense de l'Environnement au Burundi, which seek to involve communities and individuals in reforestation efforts. Partnering with local NGOs undertaking participatory afforestation programmes would help to expand their efforts to emergency settlements programmes, thus fostering landscape resilience and creating employment opportunities around newly built shelters and settlements through the offset of material impacts and local tree planting.

Collaborative approaches between shelter and environmental programmes could help to foster better-informed responses about environmentally sustainable materials, to bolster markets for new less-impactful materials, and could help to bring the resilience ecosystem benefits of forests closer to the settlement programmes themselves.

CHALLENGES AND SUCCESSES OF LOCAL PROJECTS: COMPARING THE TRADITIONAL DEVELOPMENT APPROACH WITH COMMUNITY-BASED DEVELOPMENT IN TANZANIA

Daniel Paul

Introduction

Engineering development interventions have traditionally aimed at improving large-scale infrastructure, often focusing on construction and civil engineering projects. These projects will often seek to achieve large-scale development to water, power and transport systems. However, there is growing research that supports the view that smaller, more localised projects have equally as impressive results. This has grown the field of humanitarian engineering, bridging the gap between aid and development, leading to the concepts of community-based (or community-driven) development (e.g. see Dongier et al., 2003; Mansuri and Rao, 2004). In such projects, skilled and trained engineers from developed countries can provide support to local communities in developing countries, enabling those communities to overcome issues they lack expertise in, improving the intervention, its impact, its sustainability and its appropriateness.

Although community-based development projects can have positive impacts and improve the lives of those they support, the implementation of these is more complex than larger-scale, traditional civil engineering projects. Success on the former needs a much broader skill set, necessitates greater flexibility, and requires engineers with strong problem-solving abilities.

That is not to say traditional engineering projects in developing countries are straightforward, however. Complexities arise in working practice, dealing with various stakeholders and ensuring project sustainability.

This sub-chapter seeks to analyse the challenges encountered when conducting traditional civil engineering projects in comparison with more localised community-based development projects. The author presents their own experiences managing two contrasting projects, combining practical experience with academic hindsight. Although two specific projects are used to highlight some learning points, the general discussion incorporates managing numerous humanitarian engineering projects from 2011 to 2015. The projects in question both took place in north Tanzania, and provide learning on the skillset needed of engineers conducting similar projects in the future.

Comparison against the project management process creates discussion on issues around initiation, planning and execution. A critical analysis is provided on the impact of the two projects as well as their sustainability. Finally, lessons learnt from a management perspective are provided, focusing on the unique skillset each type of project asks of from humanitarian engineers.

Background to community-based engineering projects

Engineering for development often focuses on civil and structural engineering interventions in developing countries. Projects such as the Helmand Arghandab Valley Authority irrigation project or the Rajasthan Canal Command Area Development Project require engineers with technical knowledge and project management experience. These projects often have large and tangible effects on a country's development, providing benefit to large numbers of the local population (Cruickshank, 2013).

It is well documented, however, that more localised, community-based development projects can have equally impressive effects and spur grassroots growth (Dongier et al., 2003). Athough the geographic area they take place in is often small, the impacts can be much wider than the project's original 'zone of control'; this is a contrast to traditional projects, which have a large zone of control but a narrower impact (Cruickshank, 2013). Centrally led projects are usually aimed at meeting national objectives, often neglecting local communities' true needs and being less inclusive (Casey et al., 2011). Interventions led by the local community have the possibility of being more 'responsive to demands, more inclusive, more sustainable, and more cost-effective' when compared with centrally led, traditional interventions (Dongier et al., 2003).

Although local communities often create localised solutions to their issues, humanitarian engineers can bring knowledge of best practice from post-industrial countries, allowing local communities to benefit from subject matter expertise they often would not have access to (Ali, 2015). They can supplement work carried out by humanitarian aid (provision of food, clothes, shelter, etc.) by improving livelihoods and access to healthcare, services or infrastructure, and facilitate a smooth relief/development transition (Cruickshank, 2013). Humanitarian engineers set themselves apart from aid workers by using science and technical knowledge to meet the basic needs of a population, rather than the delivery of goods or services (Mitcham and Munoz, 2010).

There are shortcomings to humanitarian engineers being involved with both traditional and community-based projects. Easterly (2013) discusses the issues of outsider experts trying to create solutions for the poor in general. Such views are reinforced by Ramalingam (2013) discussing the international aid sector, which uses antiquated top-down models to overcome issues, without considering the true nature of the system, or community, they are conducted in. Such interventions often fail and are ineffective or unsustainable when outsider help inevitably leaves.

It becomes the role of the humanitarian engineer to not only provide technical know-how to solve an issue but also do so in a way that allows the project to be self-sustaining when the engineer leaves (Amadei *et al.*, 2009). Humanitarian engineers must take knowledge learnt in the global north (or techno-bureaucratic knowledge) and apply this to community practices (or local–traditional knowledge) (Ali, 2015). In doing so, humanitarian engineers are not only able to help produce effective interventions but also ensure their sustainability (Amadei and Wallace, 2009). The two following case studies discuss how this is done over the project management process.

Project backgrounds

A map showing the locations of the projects is given in Figure 4.7.

Monduli – district hospital water tank and rainwater harvesting system

The traditional civil engineering project that will be discussed is the construction of the water system serving the new wing of Monduli District Hospital. The project aim was to ensure an adequate water supply for the hospital's two new wards, collecting water during the rainy

Figure 4.7 Map of the project locations: Monduli and Ng'uni, Tanzania

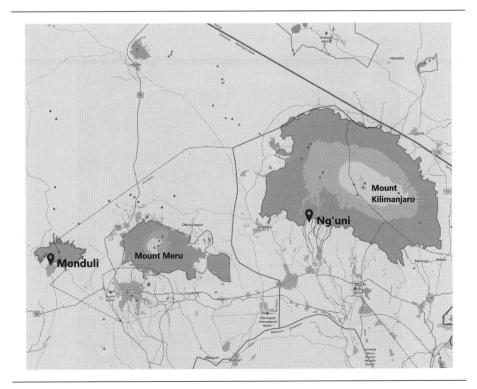

seasons (March to May, and November to December) through a rainwater-harvesting system (consisting of guttering and cleaning system) attached to the wards, and storage of water through the mains supply during the dry months (January to February, and June to October). This would allow the wards to have consistent access to water, regardless of season or mains supply.

The project took place from August 2014 through to December 2014, with much of the construction phase occurring in September. It included the construction of a 50 000 l water tank to support the new ward, which was under construction, and the installation of the rainwater-harvesting system. Planning for the project took place from July to August, with the digging of the foundations and construction of the brickwork occurring the following month (Figure 4.8). Installation of the guttering and associated mechanisms to make up the rainwater-harvesting system, as well as the completion of the water tank, occurred through October and November. Much of the work on the project was conducted by a local construction firm, which provided manual labour, with specialist engineering support from both local and two international engineers. The project was completed in time for the new wards opening in December 2014. The project was funded by the Tanzanian Government and the UNDP.

Ng'uni – Nure Women's Dairy Cooperative

The community-based development project that will be discussed is the reinforcement of the road running adjacent to the dairy cooperative buildings, initiated by the Nure Women's Dairy Cooperative, in Ng'uni village. The dairy cooperative has been a long-running initiative in the

Figure 4.8 Water tank under construction

village, buying surplus milk from local farmers and selling this on at a profit at the local market in Moshi. It expanded its operation in 2012, to purchase a refrigerator to store milk and make fewer trips to market, therefore reducing overhead costs. The road reinforcement aim was to reduce the likelihood of the road being damaged in heavy rains, which would restrict access to the market for 3–5 months of the year.

The project took place in September 2014, and consisted of the reinforcing of the road running from the centre of Ng'uni, along Nure Ridge, up to the edge of Kilimanjaro National Park. The construction included levelling the road; laying sand, murram and rock aggregate; compacting this into the soil; and identifying weak points in the road prone to landslides, and reinforcing these with concrete. The project was guided by two international engineers, with much of the work conducted by the local community and a small contingent of international volunteers who were teaching at a nearby school. The project was funded in part by Cwlwm Monduli, a small Welsh charity, and the local community, including the dairy cooperative itself. Although the project was initiated by the cooperative, funding from other community businesses and organisations was secured by the cooperative's chairwoman after selling the wider benefits of reinforcing the road – greater access to markets, improved access to healthcare and education, and the possibility of increasing tourism through the village, as the road led to an old and no longer used entrance to Kilimanjaro National Park.

Project management

The two projects will be examined along the first three phases of the project management process – initiation, planning and execution (Figure 4.9) – as well as in relation to project management theory in general. Although there were clear differences between the two projects

Figure 4.9 The project management process. (Adapted from Westland, 2006)

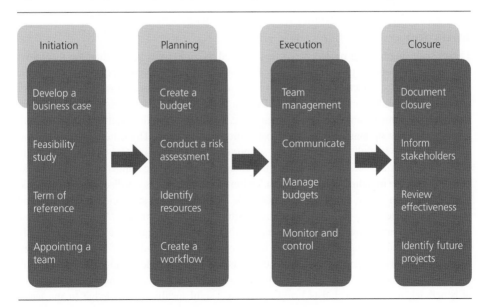

Initiation	Planning	Execution	Closure
Develop a business case	Create a budget	Team management	Document closure
Feasibility study	Conduct a risk assessment	Communicate	Inform stakeholders
Term of reference	Identify resources	Manage budgets	Review effectiveness
Appointing a team	Create a workflow	Monitor and control	Identify future projects

Table 4.1 Comparison of the two projects across the project management process

	Monduli water tank project	Nure Women's Dairy Cooperative
Initiation	■ Little control over selecting projects ■ Projects aligned to higher-level/national objectives ■ Input came in assembling a team	■ Large role played in selecting projects ■ Greater interaction with local stakeholders ■ Needed to select the most appropriate interventions
Planning	■ Planning was time consuming ■ Meetings were often formalities ■ Planning was meticulous and in greater detail ■ Planning was siloed and needed to be 'stitched' together by a humanitarian engineer	■ Planning was shorter and more straightforward ■ Planning was not a natural action; there was a want to 'do' straightaway ■ The community was involved throughout
Execution	■ Markets were at an increased risk of external shocks ■ Labour needed to be used wisely to get maximum cost/benefit	■ Less detailed planning had a negative impact when issues arise ■ The community was involved with the project as a whole

in the initiation and planning stages, the difficulties encountered in the execution stage were similar and more related to conducting projects in a developing country.

Both projects presented unique learning points that, when combined, allow the humanitarian engineer to better prepare for future projects. A summary of these learning points is provided in Table 4.1.

Initiation

Larger-scale engineering projects are often dictated by authorities higher than the local government (e.g. a national government or even regional/intra-state governance bodies such as the UN). In this way, projects have a top-down nature and the business case is more aligned to a national objective (Msangi *et al.*, 2014). From a project management perspective, engineers have limited input in developing a business case, where they would usually assess the feasibility of different options or select the most appropriate solution. The water tank project was selected by the Monduli local authorities, supported by the Tanzanian Government in a bid to increase access to drinking water across the country. The only area under the control of the author during the initiation stage was the selection of team members. As discussed by Wysocki (2013), selection of the appropriate 'super team' is an important step, and requires a sound knowledge of the likely technical specialists required during the project. Although there is a tendency for international humanitarian engineering project managers to select predominantly international staff, who generally have a more complex understanding of international engineering standards and practices, this can create issues in the development context. Although these projects need to meet set construction standards and the international experience is beneficial, national specialists bring extremely useful localised knowledge to the team.

They have a greater understanding of building techniques, material suppliers and the quality of building materials. The latter cannot be overlooked in developing countries, where regulations are less stringently followed or monitored. This became apparent during the project, when a local engineer noticed issues with the quality of the blocks required to build the sidewalls of the tank. Not only could they notice quality issues but they were also able to go to the local supplier, select higher-quality bricks and have them supplied to the site the next day. International engineers without knowledge of local building materials would have likely missed such an issue, nor would they have had the local knowledge to organise an alternative supply in such a short time. This was experienced on several other projects, where the quality of materials was either taken as granted, or the negotiating power of foreign engineers was undermined when dealing with local merchants.

For community-based projects, humanitarian engineers have a larger role to play in the Initiation stage. Greater interaction and communication with the local community are needed, with time spent understanding the issues faced, the proposed local solutions, and how these can be better engineered to ensure their appropriateness and sustainability. Engineers working on community projects often lack the accountability mechanisms that larger projects often have; this can cause them to skip spending time understanding the needs of the community in favour of implementing sooner (Amadei and Wallace, 2009). This can result in solutions that work in principle, but commonly fail once the engineer leaves; a lack of local knowledge to maintain solutions, inappropriate use due to lack of knowledge or sub-optimal construction techniques fail to create appropriate solutions. In this regard, the engineer's role is not to pose a solution but to understand what the local solution is and find ways to improve upon this (Schumacher 2003). This often takes much longer than simply identifying a workable solution, as is often done in traditional projects.

Working with the community is not always straightforward, with the community itself not always able to articulate the issues it faces or being unaware of appropriate solutions (Msangi *et al.*, 2014). This was evident in Nure, with the community proposing tarmacking the road after such solutions had been implemented on the road between Moshi and Arusha. The lack of maintenance and inappropriate initial surface would have made this solution fail after a short time. Furthermore, tarmacking the road would have considerable cost implications and would have restricted the irrigation channels that flow across many parts of the road. The alternative solution of using sand, murram and rock aggregate was ultimately more appropriate in the circumstances. Amadei *et al.* (2009) stated that closely working with the community to understand its needs allows engineers to devise 'long-lasting, successful solutions that are respectful of the community itself, its people, and its environment'. Although it took multiple stakeholder meetings, days surveying, several iterations to the proposals and ensuring buy-in from multiple key stakeholders, such time spent ensured greater buy-in and more sustainable impact from the project.

Planning

Traditional and community projects differ greatly in the planning stage. While traditional civil engineering projects often require lengthy planning meetings, well-documented workflow charts and detailed budget sheets, community projects are much more laid back. There are advantages and disadvantages to this. The lack of detailed plans for community-based projects means issues are often not identified ahead of time, with the humanitarian engineer having to react to issues as they arise and find solutions on the spot. The planning associated with traditional projects means that such issues are often identified through risk assessments and project plans, with appropriate budgets attached to solve these should they arise.

However, the need to meticulously plan can hinder traditional projects. In the author's experience of working on the water tank project, large portions of time were spent in meetings with different contractors and stakeholders. Although many of these were necessary, such as meeting suppliers and contractors, many were also formalities and resulted in few tangible outputs. In such cases, meetings were held to introduce the project and brief the stakeholders on the main points. These were then followed up with further meetings where actual planning took place; such meetings would be focused on specific aspects rather than the project in general, requiring multiple follow-up meetings. Although there are warrants to this, namely each individual stage of the project being well planned, the process is time consuming and tedious. In this regard, however, planning is siloed. The stakeholders are engaged with individually, rather than as a whole. It then becomes the humanitarian engineer's job to ensure all of the plans work together and that timelines match up.

With community-based projects, there is a tendency to 'do' as soon as possible. The role of the humanitarian engineer then becomes slowing the community down so that appropriate planning can take place, especially when identifying budgets and the likely costing of materials. Localised projects do have one benefit here – community members can leverage favour with local suppliers to reduce the cost of resources. This was particularly useful in Ng'uni, with the supplier of the road materials meeting 25% of the overall cost, as he grew up in village and still had family living there. He also agreed to pay the wages of the drivers and fuel, removing the delivery cost for the community, therefore reducing a large proportion of the overall cost.

Unlike the experiences encountered with the water tank, planning for the road reinforcement was relatively straightforward. Planning sessions took place in the local school, and included key stakeholders from the village: the village elder, the local pastor, the chair of the Nure Women's Dairy Cooperative, a local community group leader, a former headteacher of the local secondary school, the current headteacher of the primary school and a local engineer. Such a committee with a diverse stakeholder make-up was not unusual; this group represented the community as a whole, and allowed diverse thinking on how projects could be completed. It also meant the project was signed off and agreed upon by the social hierarchy. In this project, this committee was critical in both increasing financial support as well as mobilising the community to assist. This meeting was then followed by one short meeting with the local materials supplier, facilitated by a member of the community. This streamlined and holistic approach was more efficient than experiences with the water tank project.

Execution

Execution of projects in developing countries can often be difficult in general, regardless of their size. Unplanned shortages of resources, and labour strikes, as well as security and safety risks to staff, all hamper efforts. Furthermore, differences in work practices, techniques and management can be frustrating when trying to manage a project with engineers who are used to practising in developed countries.

A case in point to highlight the unplanned difficulty of managing projects in developing countries was the laying of resources for the road project. The planned approach was to have tipper trucks bring up sand, murram and crushed rocks in that order. The sand and murram were to be dropped centrally in four strategic locations, then transported to the areas of road where they were needed. The rocks were to be dropped at the area of road where they were required, reducing the burden of transporting these. The rocks selected were larger than needed, and were to be broken on site; this option was cheaper and had less logistical burden. However, as the truck drivers were paid per load, there was competition between drivers to drop as many loads as possible. The first truck that arrived dropped its load of rocks at the start of the Nure Ridge, blocking the road for the other trucks (Figure 4.10). This caused long delays for the other lorries and for trade out of the village until the boulders could be broken up and moved.

The lack of structure and planning, as well as the informal nature of the labour, made coordinating the drops difficult. Local drivers had been given little information about the project and had not been introduced to the project manager (the author) nor any of the community

Figure 4.10 Map of Nure and the proposed drop sites versus the first drop blocking the road

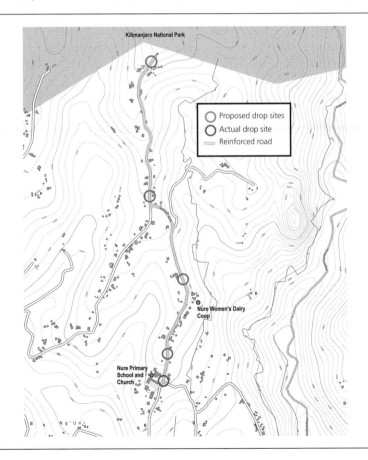

leadership. For this reason, they did not listen to directions, dropped loads at unplanned points and did not follow coordination when difficulties arose – for instance, a tipper truck getting stuck in soft soil.

Likewise, issues arose during the construction of the water system. After the construction of the tank walls, there was a shortage of tanking slurry (a waterproof cement used to line the tank walls). This shortage arose due to the material supplier's cement provider cancelling an order. Although the solution to this was relatively straightforward (ordering the tanking slurry directly through a new supplier), the cost was higher than planned, which required further stakeholder meetings to ensure adequate funding, which delayed the finishing of the tank walls by 2 weeks. The knock-on effects of this resulted in delays installing the piping and finishing the tank roof. The management challenge here, besides ensuring adequate budgets, is allocating productive tasks to the staff, who are already hired and paid for. Their work must support the project's future steps, without creating complications by rushing ahead. This requires flexibility for the humanitarian engineer and the ability to think quickly to rearrange work schedules.

General management of projects

Beside the lessons learnt across the project management process, there are general lessons to be learnt from the two different projects. There are issues around timelines, the key stakeholders interacted with, the provision and management of labour, materials used and the general management of the project. Key observations on both are listed in Table 4.2.

Further to the above observations, there are general issues which face the humanitarian engineer when managing projects: language barriers with local staff and cultural differences, as well as factors as simple as the environment. An aspect that could easily be overlooked prior to deploying is the safety and security aspects. While working in Monduli and travelling to Arusha, there were two notable periods of civil unrest in which unwanted attention was drawn to the international staff as foreigners. Furthermore, there were multiple instances of petty crime, such as pickpocketing and harassment. While such issues did not occur in Ng'uni, one concerning aspect was the distance to adequate medical care – roughly 4 h by road, depending on the reliability of transport. An injury as simple as a fall can become life threatening. Such considerations in planning effective security risk assessments and appropriate medical plans require as much detail as design schematics and project plans, necessitating a very detail-oriented mindset from the humanitarian engineer.

Project impacts

The impacts of the Monduli water tank project were easy to map and predictable: the system allowed the supply of water to the newly constructed ward and storage of 50 000 l of water. However, the rainwater-harvesting system did not have huge benefits in the years following (2015–2018) due to below-average rainfalls in the region (World Weather Online, 2021a), meaning there was a greater reliance on filling the tank from the mains supply. Furthermore, the construction of the water system only provided benefit to the hospital ward, meaning its impact was very narrow.

Table 4.2 Comparison of general management differences between the two projects

	Monduli water tank project	Nure Women's Dairy Cooperative
Timeline	■ Initial meetings had no tangible output ■ Initial planning was time consuming ■ Incentive for contractors to spend as long on the project as possible ■ Timeline dependent on contractors and other projects	■ Meetings ended in tasks, expediting the process ■ Less time spent planning ■ Incentive for the community to complete the project quickly ■ Timeline was dependent on village events (e.g. church, weddings and funerals)
Key interactions	■ Meetings were with higher-level managers, not the implementing partners ■ Stakeholders were not engaged ■ Government representatives took a keen interest	■ Stakeholders were part of the whole process, from planning to execution ■ Greater interaction with the community in general ■ Less interaction with the government, but keen interest from the local leadership
Labour	■ Labour was skilled, trained and appropriately equipped ■ Support from specialists where needed ■ Labour worked within set hours	■ Labour was often unskilled, although experienced with manual labour (e.g. local farmers) ■ Specialist labour was non-existent ■ Labour continued working until the job was done
Materials	■ The water system and ward were constructed by two different contractors, creating competition for resources ■ Weak supply chains created an inherent weakness when ordering supplies ■ Cost was higher due to lack of local negotiation power/perception of wealth	■ Local networks were used to buy material at a discounted price ■ Knowledge of the most appropriate material vendors limited to community networks ■ Supply chain weakness still present
Management	■ Elements of power play with local contractors ■ Staff given clear jobs and completed them with minimal supervision ■ Management involved tracking budgets, liaising with other projects and ensuring adherence to timelines	■ Local community members followed set cultural hierarchies ■ Local community members needed close management ■ Management involved micromanaging the local community, with less focus on the budget or fitting into other projects

Interestingly, the top-down approach seen in larger-scale infrastructure projects that are dictated or directed by governments is seen to be less effective at meeting the community's needs. In discussions with the key community stakeholders during the construction of the water tank and system, it was noted by several members, including the hospital director himself, that water shortages are a common issue across the region that need a larger investment to overcome rather than a localised solution. Although the water system was a critical need for the expansion of the hospital, itself addressing inadequate access to healthcare, the more appropriate solution would be to improve water access across the region in general. The top-down approach to budgeting often neglects the true issues that locals face; devolving financial management to local government and improving community 'ownership' was a solution posed by Msangi *et al.* (2014) to address true needs at the grassroots level. Such a solution overcomes the issue of government grants being stipulated to meet national priorities – a methodology that is not always effective.

The impacts of the road construction in Ng'uni present some interesting findings. The initial aim, for the Women's Dairy Cooperative to have more stable access to market, was achieved. Peaks in rainfall in 2016 and 2018 created conditions that would have previously resulted in collapse of the road and prevented the dairy cooperative accessing the market in Moshi (World Weather Online, 2021b). As well as meeting its initial aim, there were numerous further benefits of the project: greater access for farmers to bring their milk to the cooperative, and better access for vehicles meant the women could sell more milk at the market in Moshi, increasing their profits. This in turn allowed the cooperative to buy a more reliable refrigerator to store the milk, as well as to invest in other equipment to make cheese, a product that could be kept for longer as well as be sold for a higher profit than milk alone.

Further to solely benefiting the dairy cooperative, the reinforcement of the road opened the higher parts of Ng'uni village up during the rainy seasons. It was estimated that of the 7000 residents of Nguni, over 2000 lived in the higher areas at the top of Nure Ridge. The increase in access to lower parts of the village was documented in anecdotal evidence of a higher attendance in church during the rainy seasons; this shows that access to the central part of the village was higher. This improved access to the central hub of the village, which included the local shop, church, Sunday market and the medical centre. Such an increase in access would no doubt improve the overall quality of life.

The increased trade for the village due to the road being reinforced meant that villagers increased their profits from sales of crops, milk and other goods. The success of the project and increase in trade meant local businesses turned an increased profit. This inspired the chairwoman of the dairy cooperative to rally business owners to fund a community project. The community decided on funding an IT suite attached to the church. This would primarily be used to improve education for children, with IT not being mandatory in the Tanzanian education syllabus. Local businesses would also have access to the computers for their own use, but with the overall aim for them to hire school leavers to fill more technical (IT) roles. The chairwoman stated that 'children in Nure community need the skills to operate in the Global Community'. The classroom was constructed and finished in 2016, with computers being purchased the same year with support from the charity Cwlwm Monduli.

Project sustainability

It is difficult to predict project sustainability after such a short time. Some inferences can be drawn from the costs needed by the projects after their implementation, however, as well as the engineering methods employed. Analysis of the funding and similar projects in the area can also aid in assessing the sustainability of the two projects.

Between the completion of the Monduli water tank project in December 2014 and December 2015, additional costs were sought to repair a section of guttering of the rainwater-harvesting system. Heavy winds from Mt Meru are a common problem, with the rear of the ward facing the westward slope of the mountain, causing damage to several sections. However, funding ceased at the end of the project, with the Tanzanian Government and UNDP providing no further budget for maintenance. The repair was therefore funded by the hospital itself; although this was always the intended process, lack of funding meant the unforeseen cost was a burden for the hospital, which detracted from other projects on the grounds. However, the hospital receives national funding, and therefore always has money available for general maintenance. Well-budgeted finances would mean there is always money available for repairs, or to hire local staff to carry out maintenance throughout the year. As such, the project has the potential to be sustainable in the long run.

This is contrasted with the road project, which sought no extra funding. Repair work did need to be carried out on several parts, however, which highlights the issues of using unskilled labour. The costs for this came from communal funds and local businesses, aware of the benefit the road reinforcement had provided to their trade. This is not to assume the road project is more sustainable. The repairs were minor and required little funding. Large shocks to the road, such as extreme weather events, could cause greater damage, which is harder and more costly to repair. The risk from climate change on landslide risk is well documented (Gariano and Guzetti, 2016; Lee, 2017). Such a change in weather patterns can be seen in the Hai district already. The peak rainfall for 2014 was 85.96 mm during the second rainy season, typically the wetter of the two. This is contrasted with 2020, where there was a peak rainfall of 195.7 mm in January alone – a month that is typically a dry season (World Weather Online, 2021b). Furthermore, examples of other road projects in neighbouring villages also show degradation over time. These roads often exist further outside a community, where ownership is not as obvious, and the backgrounds of these are unknown. Such roads might not have had the community involved throughout the project management process, therefore lacking local ownership. It is obvious, though, that the construction methods used are not sustainable without regular upkeep and maintenance. Although this is possible for the villagers in Ng'uni to do, it requires both labour and materials, as well as money for the latter. This depends largely on their ability to access local markets to make profits, which in turn is dependent on market prices for what they are selling. Should market prices drop, less money is available to maintain the road, which in turn makes it more susceptible to damage from extreme climate events, thus leading to local traders not being able to access markets and make money, further reducing the funds available to make repairs to the road. Such a fragile relationship exists, threatening the sustainability of the project.

Although this brief analysis could bring one to believe that traditional projects are more sustainable, this is not true. Both projects are reliant on money – the water system on external funding from the government, the road on money from within the community. Neither is guaranteed.

Lessons learnt

The skills needed on the two projects differ. Larger-scale development projects require more traditional management skills, where issues are around contract negotiations, ensuring fair prices for supplies, and that individual timelines on the project match up so that the whole is completed on time. On the other hand, community-based projects require a more people-centred skillset: being able to run community meetings, where English is often not spoken widely; understand the broader needs rather than the specific needs of strong characters; and ensuring community buy-in. Community-based projects require the humanitarian engineer to take a wide range of knowledge of global engineering practices, understand the local solutions, and bring the two together to improve methods and increase sustainability. Table 4.3 highlights the differences in the skills necessary between the two types of projects, as well as the general skills needed to be successful on both.

Table 4.3 Skills needed on traditional and community-based development projects

Skill Needed	Example
Traditional project	
Negotiation skills	Agreeing on contracts with suppliers, negotiating the prices of supplies
Project management skills	Ensuring different elements of projects finish at the same time, matching timelines
Budgeting/financial planning	Ensuring supplies are within budget, keeping budgets aside for likely issues, payment of staff and contractors
Formal risk assessments	Assessing risks to labourers, identifying possible risks to markets
Community-based project	
Needs assessment	Identifying the needs of the community, working out what affects the whole community
Cultural awareness	Ensuring appropriate interactions with community members
Communicating with locals	Organising village meetings to collect information, communicating technical problems to local villagers, organising villagers during the project
Problem-solving	Overcoming problems as they arise, applying broad engineering solutions to local practices
Working in developing countries	
Cultural sensitivity	Paying proper respect, dressing appropriately, understanding formalities and societal hierarchies
Safety and security consciousness	Understanding the local risk, being aware of dangers to yourself/your team, acting in a security-conscious manner
Adaptability	Being prepared for unplanned issues, knowing alternative solutions, keeping plans flexible
Personal effectiveness	Staying hydrated, eating well, being fit enough to deal with the stress of the climate/environment

Conclusions

Although traditional engineering projects have wide-ranging benefits and often have greater funding, this sub-chapter has discussed the warrants of more localised, community-based development projects. These projects can address issues facing the community with more-detailed knowledge of its true needs. Such projects require humanitarian engineers to adapt technical knowledge learnt in developed countries to local practices. The skills needed between the two discussed projects contrast as well: traditional projects require project management skills and the ability to budget, while community-based development projects require people skills and the ability to solve problems as they arise.

The impacts of the two projects are interesting to assess. While the water project was effective at meeting the project aims, the impacts were narrow in focus: water was supplied to the hospital. The impacts of the road project were much further reaching. The dependability of the road allowed an increase in trade; the surplus profit from this was then put into creating an IT classroom, which benefited children in the village.

The sustainability of both projects is hard to judge. However, the effect of climate change and extreme weather patterns, as well as the general strength of the economy, threatens both. For the water system, lower than average rainfalls make the rainwater harvesting system redundant; for the road project, a higher than average rainfall threatens the road stability. The community-based problem had greater ownership of the project, although it is more susceptible to shocks – minor repairs were funded and carried out by the local community, but larger issues, especially those likely to be experienced due to extreme weather events, would have wider ramifications and be harder to recover from.

This sub-chapter provides learning for future humanitarian engineers conducting localised projects, especially around the skillset necessary and the way in which projects are managed. Although each project will be different and present a range of challenges, adaptability and the ability to quickly problem-solve will hold any engineer in good stead.

UNPACKING DENSITY IN SITES – OBSERVATIONS FROM THE SITE-PLANNING GBV PROJECT IN MAIDUGURI, NIGERIA

Joseph Ashmore, Jim Kennedy, Alberto Piccioli and Amina Saoudi

Introduction

Every year, humanitarian shelter organisations assist far in excess of 10 million people affected by crises. In 2019, 79.5 million people were forcibly displaced, many having been displaced for decades (UNHCR, 2019). During the year, nearly 25 million people were newly displaced by natural disasters and 8.5 million people were newly displaced by conflict (Internal Displacement Monitoring Centre, 2021). These numbers include both those who were displaced within their own countries as internally displaced people, as well as the many others who were forcibly displaced across international borders, becoming refugees.

The majority of those who have been displaced find shelter with host families, host communities, or – if their own resources permit – are able to rent houses or apartments until a more durable option becomes available. Living in sites (for the purposes of this sub-chapter, the word 'site' refers generally to sites, camps, spontaneous settlements, or other contiguous groupings of shelters and non-shelter facilities, planned or unplanned, for those who have been displaced from their homes by conflict or natural disaster) is widely acknowledged to be an option of last resort, but for a percentage of displaced people, it is an unfortunate reality. Sites – whether camps fully planned by governments or humanitarian organisations, or more usually spontaneous settlements created by the families living in them – are often the location of heightened risk of violence, may isolate those living in the sites from the re-establishment of economic independence, and can have a concentrated negative impact upon the surrounding environment.

In those circumstances where sites are the only viable shelter solution, they are ideally planned and constructed in advance of people arriving. However, the reality is that, in many places, people arrive at locations and site planning has to follow. For any planning activities within a site, density is a key metric.

The Sphere Handbook (Sphere, 2018) establishes the following indicators – which unfortunately are seldom attained

- 45 square metres for each person in camp-type settlements, including household plots
- 30 square metres for each person, including household plots, where communal services can be provided outside the planned settlement area.

And over-density in a site population can be a matter of life and death. High population density in sites has been linked to risks of catastrophic fire, blocking of evacuation routes and increased demands for clean water. These risks threaten minimum humanitarian standards and disrupt access to life-saving services. In addition, high population density may significantly increase the exposure, particularly for women and girls, to acts of gender-based violence (GBV).

In response to these issues, a project was piloted in Maiduguri in northern Nigeria during 2018 and 2019 to make assessments, followed up with action plans to address issues of population density. In so doing, it found errors with previous assertions that high-density populations alone affect GBV risks and perceptions of safety.

While many of the common technical guidance documents used by site planners provide a framework of minimum spatial standards (derived in turn from some of the mass housing-engineering concepts of the first half of the 20th century, originating in the 1920s Frankfurt School of minimum *existenz* low-income housing, to 1960s Latin American 'sites-and-services' rapid-expansion housing settlements), in order to assure the minimum distances needed between shelters to provide safety and dignity for everyone in the site, the reality is that many sites do not meet these standards, either in part or in whole. In the majority of cases, site planners and engineers only arrive at the site once it has already come into existence, established spontaneously by those who have lost housing or been displaced, and without any overarching organisation or equitable distribution of the available land. In many other cases, even if the site had been initially planned according to globally accepted minimum standards, intervening years of a protracted humanitarian crisis may have resulted in increases in the population, alongside appropriation or encroachment into shared spaced in sites by those needing to expand their shelters, or looking to establish livelihoods and self-reliance in their site existence.

The relationship between high population density, health risks and protection in sites is noted in a number of guidance documents (Global Shelter Cluster, 2013; UNHCR, 2021). And the IASC (2015) has stated

> Overcrowding in urban areas or camp situations can exacerbate family tensions, which in turn can contribute to intimate partner violence and other forms of domestic violence. Overcrowding can also increase the risk of sexual assault by non-family members, particularly in multifamily tents, multi-household dwellings or large communal spaces.

However, the relationship between GBV risks in low-density areas within sites or areas with significant density gender disparity has received less attention.

In conducting the field assessments, we focused on GBV risks rather than GBV incidents. Experience from the field has shown that quantifying the prevalence of GBV in humanitarian contexts is not advisable. Such efforts not only divert the limited resources for response but also misrepresent the amplitude of the problem and risk further exposure of survivors (resulting in further ostracism or retaliation). In response, these assessments focus on GBV risks as they relate to perceptions of safety, in particular for women and girls.

Methodology

Observational audit exercises were undertaken in eight sites in Maiduguri, by groups including site planners and camp management staff, staff from NGOs and the Red Cross movement, and local officials charged with camp administration. The observational audits are a relatively rapid three-step process of analysing site maps to identify likely areas with a heightened risk of GBV, doing small-group transect walks guided by a GBV-focused discussion questionnaire, and then planning sessions to identify and prioritise site-planning interventions designed to address the identified risks. The observations made in the sites had the purpose of visually identifying where in sites there was a heightened risk of GBV related to population density.

The IOM (2017) booklet *Site Planning: Guidance to Reduce the Risk of Gender-Based Violence* discusses how GBV risk may be related to population density. Examples include

- overcrowded male-dominated markets, making it easier for physical harassment or intimidation
- narrowing of pathways by the informal extension of shelters or shelter plots, potentially increasing fear and the likelihood of hidden attackers
- increased waiting time for communal latrines or showers, or isolated placement, increasing the exposure to physical attack
- longer distances and isolation when gathering firewood or food, increasing the chances of rape or sexual assault.

In Maiduguri, following observation exercises, actors within the sites were either able to suggest specific recommendations to help address GBV risks or identify the need for further investigation using other tools, such as focus-group discussions or safety audits conducted by trained staff.

There were limitations to the study. The sites that were visited in Maiduguri were relatively small, and located close to the urban centre, affecting both the social and economic dynamics of the site populations. Secondly, some sites were located within the grounds of old or unfinished civil-servant housing colonies, where many of the empty houses had been occupied by those living in the sites. This in some cases created visually clear borders for high and low population density in public areas within the sites, making it easier for stakeholders to identify and respond to the localised risks. (It should be noted, though, that this does not take into account the density of occupation of the old housing itself.) In these aspects, the eight sites are perhaps not typical of many other sites around the world, and the findings cannot be extrapolated.

Key findings

In the Maiduguri context, there were five key findings regarding population density from the observational audits, and shared across all the visited sites. These findings point to the connections between localised population density, GBV risk and other major cross-cutting issues.

1. **Density and livelihoods**. The official central markets in all the sites visited were densely populated – but dominated by men and boys, with few if any women present. Evening grill fires and shops with public music systems also gave evidence that the male dominance of the area was 24 h a day, with increasing risks of GBV at night due to alcohol and drug use. Women's and girls' perceived risks of harassment and sexual violence deterred them from working within or visiting the central markets. In contrast, women's livelihood opportunities focused on small stalls adjacent to shelters, in the shelter blocks of the sites. The location of these stalls allowed women to be near and watch their shelters, livestock and small children while also engaging in livelihood activities. On the other hand, the range of items that could be sold in these stalls was much more limited than in the main market stalls, reducing income generation and increasing dependency on higher-male-income earners. Women's and girls' exclusion from the markets was also caused by the empty no-man's lands that surrounded the central markets. Such wide-open spaces, without community vigilance, further isolated women in their shelter blocks for concern of attack.

   ▪ The Maiduguri site-planning team proposed to incrementally increase the areas where women feel safe to conduct market and livelihoods activities. This would be through better lighting; the construction of physical structures, such as stalls used by women; and monitoring of, and efforts to contain, the men-only TV cinemas, informally constructed between the main markets and the housing blocks in some sites.

2. **Density and liveable space**. Much of the daily lives of many women in the sites was constrained to the shelter block areas. These shelter blocks were physically encroached upon by a variety of different self-built extensions, and experienced little or no temporary de-densification during daytime hours. Adding extensions, such as outdoor kitchens, allowed the women to expand the basic first shelters, and also replicated cultural practices from home. However, such extensions, combined with self-built shade areas adjacent

to or between shelters, effectively blocked evacuation routes, in case of fire or threat of violence.

■ The Maiduguri site-planning team has undertaken the creation of surface-drainage channels, to reduce infectious disease, increase the amount of dry usable surface area, and increase the organisation and safety of pathways and transit routes. As most of the shelter blocks were surrounded by open space, further mitigation measures included constructing wider shade areas for communal use, and for increased market activity, extending from the edges of each shelter block. The provision of adequate market space within accessible range of the women living in the blocks would thereby reduce exclusion from broader income generation experienced in the larger market.

3. **Density and access to sanitation**. In the majority of the sites visited in Maiduguri, the latrine and shower blocks had been constructed in open space, between shelter blocks. However, despite the fact that the official latrines were in clearly accessible spaces, many were obviously unused, or effectively abandoned. Subsequently, the number of low-quality, self-built latrines, with shallow pits and placed close to the shelters, proliferated in the sites, increasing the likelihood of communicable disease. In some cases, the inferior self-built latrines were only 30–40 m from the nearest official latrine block, well within the distance range of Sphere indicators. Closer analysis highlighted that the low population density immediately surrounding the latrines negatively affected latrine use. Latrines were located in an emptied-out unoccupied space – the antithesis of having any community watchful eyes or 'defensible space', and of particular concern for women and girls. This was exacerbated by the facilities not being segregated properly, and the lack of distance between male and female facilities. In some cases, the physical isolation of latrines was also heightened by the lack of night lighting. In fact, the nearest structures to the latrines were often shops (or in some cases informal rest areas under trees) that were only frequented by men.

■ The Maiduguri site-planning team has started to divide and demarcate latrines for men or women and to provide adequate lighting for the latrines, as well as other areas in the site. Other proposals included the addition of privacy barriers to the latrine design, so that latrines could be transferred and reinstalled closer to the shelter blocks, without compromising the privacy and dignity of those using them. This has been done, although with initially mixed results, and requests that the design be done with input from the community. Other subsequent designs have been adapted with feedback indicating more success.

4. **Density and access to education**. Each of the sites visited in Maiduguri had temporary schools constructed by humanitarian organisations within the site boundaries. For the most part, schools were not located in the centres of the sites. Instead, they were placed at the edges of the sites, usually close to the front entrance. A variety of reasons, ranging from available space to the proximity of water supply, influenced the allocation of facilities within the site settings. The end results in Maiduguri were schools located at the far ends of sites – a significant distance from the shelter blocks. Girls walking to and

from school were in many cases faced with a choice of routes: walking through or along the male-dominated marketplaces, which had high and disparate population densities with perceived risks for verbal and sexual harassment; alternatively, walking through wide unoccupied spaces, where the extreme isolation provided heightened opportunities for physical and sexual assault.

■ The Maiduguri site-planning team actioned that they would include grouping other facilitates close to the schools that are supportive of the presence of women and girls. These facilities might include women's friendly spaces, health centres or women's vocational training workshops. This would extend the safe area immediately surrounding schools to wider parts of the site, increasing safe access for young women and girls. As the programme expands into larger sites outside the urban boundaries, there have been further discussions about how to balance access to all services throughout the sites, with multiple schools seen as more accessible than just one main school in large sites, with similar considerations needed for safe spaces, clinics and access to other essential services.

5. **Density and connections outside the sites**. All but one of the sites visited had populations of less than 20 000 people, and were located in allocated lands within the urban boundaries of Maiduguri. Although there were both central markets and individual stalls or kiosks within each site, these were small, and with few exceptions provided only daily goods and basic foodstuffs. For access to other markets or to wider social networks, many of those living in the sites were making frequent journeys outside the sites. At the same time, deliveries of food, and in some case potable water, were brought by hand into the sites every day, by traders from outside. Most of the sites visited in Maiduguri had a single entry/exit out of a walled compound, controlled by local army units. The entry gates were not only a bottleneck for vehicles and pedestrians but had also developed their own localised densities of populations – both stationary and flowing. These spaces often had low gender parity and experienced high levels of traffic at different times throughout the day, despite the constricted space. This bottleneck led to the entry/exit areas of the sites being identified as locations of heightened risks of verbal harassment, attack, and even trafficking, sexual exploitation and extortion. The army units, and those visiting the sites to bring goods and trade, did not actually live in the sites. Passage was given to males with influence, connections or monetary bribes, holding little or no accountability to the site population. Although those entering for trade provide access to essential goods, they too introduced heightened and various GBV risks of particular concern to women and girls, especially sexual exploitation. Other males given passage to the site are often entering for drug or alcohol use and sexual violence.

■ The Maiduguri site-planning team has ensured that sensitisation targeting camp coordination and camp management and local security within the site has been ongoing on camp rules and violence prevention for the men who work at or pass through the entrance gates. It was also planned to divide the entrance more clearly into separate lanes for vehicles and for foot traffic. This would permit greater organisation for general road safety while also reducing the opportunities for women and girls to be trapped and harassed while being squeezed into too-narrow spaces by oncoming vehicles.

Conclusions

While planned camps are to be avoided, the unfortunate reality is that far too often they are the only viable solution for sheltering displaced people. By focusing on Maiduguri, Nigeria, we highlighted that, when they do exist, their construction, upgrade and maintenance is not purely a civil engineering challenge. Instead, complex social conditions need to be understood, using a breadth of skills, including close engagement of those who will be managing the sites.

There is no one single solution that would reduce the GBV-related risks exacerbated by extremes in high or low population densities. However, the approach of identifying the types of localised density and associated risks provides a powerful, practical and inclusionary tool for mitigating GBV risks for all populations within sites.

Although the over-density of populations in sites has clear and negative effects, the experience in Maiduguri shows that the gender disparity between populations in given sites throughout sites, and even low density or isolation, also create heightened risks of GBV. These effects are more easily quantifiable, and subsequent interventions concrete, when the risk analysis is localised to specific areas within the site.

The development of a site, the changing population densities within a site, and the changing levels of GBV risks in different parts of the site need to be seen clearly as dynamic processes and not static. Considering the site as an interconnected collection of many parts offers the potential to address problems in cases where large-scale closure, expansion or replacement of a site are not options. The set of proposed actions following from the observation work in Maiduguri provides examples of some ways forwards where the engineering expertise of site planners may be combined with camp management, programming by other humanitarian actors, and partnership with those living in the sites, in order to develop and implement a palette of appropriate shelter or infrastructure insertions, to mitigate specific risks.

Acknowledgements

Jessica Izquierdo and Aliyah Sarkar (International Organization for Migration) are thanked for their help.

REFERENCES

AFF (African Forest Forum) (2011) *Forest Plantations and Woodlots in Burundi. AFF Working Paper Series*, vol. 1, issue 11. AFF, Nairobi, Kenya.

Alam SA and Starr M (2009) Deforestation and greenhouse gas emissions associated with fuelwood consumption of the brick making industry in Sudan. *Science of the Total Environment* **407(2)**: 847–852, 10.1016/j.scitotenv.2008.09.040.

Ali S (2015) Engineering in solidarity: hybridizing knowledge systems in humanitarian and international development work. *Procedia Engineering* **107**: 11–17, 10.1016/j.proeng.2015.06.053.

Amadei B and Wallace W (2009) Engineering for humanitarian development. *IEEE Technology and Society Magazine* **28(4)**: 6–15, 10.1109/MTS.2009.934940.

Amadei B, Sandekian R and Thomas E (2009) A model for sustainable humanitarian engineering projects. *Sustainability* **1**: 1087–1105, 10.3390/su1041087.

Ashmore J and Fowler J (2009) *Timber as a Construction Material in Humanitarian Operations*. UN OCHA, IFRC, CARE International, Geneva, Switzerland. https://postconflict.unep.ch/humanitarianaction/documents/02_05-04_03-07.pdf (accessed 07/02/2021).

Bradshaw C, Sodhi N, Peh K and Brook B (2007) Global evidence that deforestation amplifies flood risk and severity in the developing world. *Global Change Biology* **13**: 2379–2395. 10.1111/j.1365-2486.2007.01446.x.

Casey K, Glennerster R and Miguel E (2011) *Reshaping Institutions: Evidence on External Aid and Local Collective Action*. National Bureau of Economic Research, Cambridge MA, USA.

Cruickshank H (2013) Beyond aid: the role of engineering interventions in contributing to international development. *International Journal of Sustainability Policy and Practice* **8**: special issue.

Da Silva J (2010) *Lessons from Aceh*. Arup Disasters Emergency Committee, Practical Action, Rugby, UK.

DDC (Direction du Développement et de la Cooperation) (2016) *Programme Regional Proecco*. DDC, Bern, Switzerland. https://www.shareweb.ch/site/EI/Documents/PSD/Topics/Local%20Economic%20Development/SDC%20-%20Factsheet%20-%20PROECCO%20-%202016%20(fr).pdf (accessed 12/04/2021).

Dongier P, Van Domelen J, Ostrom E *et al.* (2003) Community-driven development. In *The Poverty Reduction Strategy Sourcebook* (Levinsohn J (ed.)), vol. 1. World Bank, Washington, DC, USA, pp. 301–331.

Easterly W (2013) *The Tyranny of Experts: Economists, Dictators, and the Forgotten Rights of the Poor*. Perseus, New York, NY, USA.

EEA (European Environment Agency) (2015) *Water-Retention Potential of Europe's Forests*. EEA, Luxembourg, Technical Report 13/2015.

Eires R, Sturm T, Camões A and Ramos L (2012) Study of a new interlocking stabilised compressed earth blocks. *Conference: Terra 2012, Lima, Peru*. https://www.researchgate.net/publication/264119160_Study_of_a_new_interlocking_stabilised_compressed_earth_blocks (accessed 12/04/2021).

Escamilla E and Habert G (2015) Global or local construction materials for post-disaster reconstruction? Sustainability assessment of twenty post-disaster shelter designs. *Building and Environment* **96**: 692–702, 10.1016/j.buildenv.2015.05.036.

French.china.org.cn (2018) Burundi: validation de normes scolaires destinées à améliorer l'accès à une 'éducation de qualité'. http://french.china.org.cn/foreign/txt/2018-05/09/content_51187367.htm (accessed 12/04/2021).

Fuwape JA (2003) *The impacts of forest industries and wood utilization on the environment*. http://www.fao.org/3/XII/0122-A2.htm (accessed 12/04/2021).

Gariano S and Guzzetti F (2016) Landslides in a changing climate. *Earth Science Reviews* **162**: 227–252, 10.1016/j.earscirev.2016.08.011.

Global Shelter Cluster (2013) *Guidance on Mainstreaming Protection in Shelter Programmes*. http://www.sheltercluster.org/sites/default/files/docs/Protection%20Matrix_v4%20120924.doc (accessed 12/04/2021).

Grozdanic L (2014) Rammed earth Muyinga library provides a bright learning experience for deaf kids in Burundi. Inhabitat. https://inhabitat.com/rammed-earth-muyinga-library-provides-a-bright-learning-experience-for-deaf-kids-in-burundi/ (accessed 12/04/2021).

IASC (Inter-Agency Standing Committee) (2015) *Guidelines for Integrating Gender-Based Violence Interventions in Humanitarian Actions: Reducing risk, Promoting Resilience and Aiding Recovery*. IASC, Geneva, Switzerland. https://interagencystandingcommittee.org/working-group/iasc-guidelines-integrating-gender-based-violence-interventions-humanitarian-action-2015 (accessed 12/04/2021).

Internal Displacement Monitoring Centre (2021) Global internal displacement database. https://www.internal-displacement.org/database/displacement-data (accessed 12/04/2021).

IOM (International Organization for Migration) (2017) *Site Planning: Guidance to Reduce the Risk of Gender-Based Violence*. IOM, Geneva, Switzerland.

Kuittinen M and Winter S (2015) Carbon footprint of transitional shelters. *International Journal of Disaster Risk Science* **6(3)**: 226–237, 10.1007/s13753-015-0067-0.

Lee C (2017) Landslide trends under extreme climate events. *Terrestrial Atmospheric and Oceanic Sciences* **28(1)**: 33–42, 10.3319/TAO.2016.05.28.01(CCA).

Mansuri G and Rao V (2004) Community-based and -driven development: a critical review. *World Bank Research Observer* **19(1)**: 1–39, 10.1093/wbro/lkh012.

Matard A, Noorullah K, Allen S *et al.* (2019) An analysis of the embodied energy and embodied carbon of refugee shelters worldwide. *International Journal of the Constructed Environment* **10(3)**: 29–54, 10.18848/2154-8587/CGP/v10i03/29-54.

Mitcham C and Munoz D (2010) *Humanitarian Engineering*. Morgan and Claypool, Williston, VT, USA.

Msangi A, Rutabingwa J, Kaiza V and Allegretti A (2014) *Community and Government: Planning Together for Climate Resilient Growth – Issues and Opportunities for Building Better Adaptive Capacity in Longido, Monduli and Ngorongoro Districts, Northern Tanzania*. International Institute for Environment and Development, London, UK.

Peñaloza D, Erlandsson M and Falk A (2016) Exploring the climate impact effects of increased use of bio-based materials in buildings. *Construction and Building Materials* **125**: 219–226, 10.1016/j.conbuildmat.2016.08.041.

ProAct Network (2009) NRC Evaluation Report, the ecological impact of refugee/returnee programmes supported by the Norwegian Refugee Council in Burundi.

Ramalingam B (2013) *Aid on the Edge of Chaos: Rethinking International Cooperation in a Complex World*. Oxford University Press, Oxford, UK.

Savary P (2011) *Élaboration de la Stratégie Sectorielle pour le Secteur de l'Énergie au Burundi. Rapport Final Provisoire*. IED, EUEO, Ministere de l'Energie et des Mines, Burundi. http://www.euei-pdf.org/sites/default/files/field_publication_file/EUEI_PDF_Burundi_Strat%C3%A9gie_sectorielle_Jan_2011_FR.pdf (accessed 12/04/2021).

Schumacher E (2003) *Small is Working: Technology for Poverty Reduction*. United Nations Educational, Scientific and Cultural Organisation, Paris, France.

Sfeir-Younis A (1986) *Soil Conservation in Developing Countries*. World Bank, Washington, DC, USA.

Skat Consulting (2017) *PROECCO Project, Innovations in Brick: Semi-industrial Brick Manufacturing*. Swiss Agency for Development and Cooperation, Bern, Switzerland. http://greengrowth.rw/wp-content/uploads/2017/12/Innovations-in-Brick.pdf (accessed 12/04/2021).

Skiadaresis G, Schwarz J and Bauhus Jürgen (2019) Groundwater extraction in floodplain forests reduces radial growth and increases summer drought sensitivity of pedunculate oak trees. *Frontiers in Forests and Global Change* **2**: 5, 10.3389/ffgc.2019.00005.

Sphere (2018) *The Sphere Handbook: Humanitarian Charter Minimum Standards in Humanitarian Response*, 2018 edn. Sphere, Geneva, Switzerland.

Stokes A, Norris J, Beek L *et al.* (2008) How vegetation reinforces soil on slopes. In *Slope Stability and Erosion Control: Ecotechnological Solutions* (Norris JE, Stokes A, Mickovski SB *et al* (eds)), Springer, Dordrecht, Germany, pp. 65–118.

Tererai F, Gaertner M, Jacobs S and Richardson D (2013) Eucalyptus invasions in riparian forests: Effects on native vegetation community diversity, stand structure and composition. *Forest Ecology and Management* **297**: 84–93, 10.1016/j.foreco.2013.02.016.

UN DESA (United Nations Department of Economic and Social Affairs) (2019) *Global Forest Goals and Targets of the UN Strategic Plan for Forests 2030*. UN, New York, NY, USA. https://www.un.org/esa/forests/wp-content/uploads/2019/04/Global-Forest-Goals-booklet-Apr-2019.pdf (accessed 12/04/2021).

UNEP (United Nations Environment Programme) (2010) *GEO Haiti: State of the Environment Report 2010*. UNEP, Port-au-Prince, Haiti. https://postconflict.unep.ch/publications/Haiti/GEO_Haiti_EN_2010.pdf (accessed 12/04/2021).

UNFCCC (United Nations Framework Convention on Climate Change) (2014) *Project Design Document: BQS Improved Cookstoves for Burundi's Schools*. UN, New York, NY, USA. https://cdm.unfccc.int/UserManagement/FileStorage/W8C3QPZ19OUHMDVG025ALNXY7BTIRF (accessed 12/04/2021).

UNHCR (United Nations High Commissioner for Refugees) (2019) Global trends: forced displacement in 2019. https://www.unhcr.org/globaltrends2019/ (accessed 12/04/2021).

UNHCR (2021) Emergency handbook: site planning for camps. https://emergency.unhcr.org/entry/35943/site-planning-for-camps (accessed 12/04/2021).

Westland J (2006) *The Project Management Lifecycle: A Complete Step-by-Step Methodology for Initiating, Planning, Executing and Closing a Project Successfully*. Kogan Page, London, UK.

Win Thin L (2018) Plant trees in refugee camps to stop forest loss and conflict, U.N. says. Reuters. https://www.reuters.com/article/us-global-refugees-deforestation/plant-trees-in-refugee-camps-to-stop-forest-loss-and-conflict-u-n-says-idUSKBN1JG323 (accessed 12/04/2021).

World Bank (2018) *Republic of Burundi: Addressing Fragility and Demographic Challenges to Reduce Poverty and Boost Sustainable Growth. Systematic Country Diagnostic*. World Bank, Newy York, NY, USA, Report 122549-BI. http://documents1.worldbank.org/curated/en/655671529960055982/pdf/Burundi-SCD-final-06212018.pdf (accessed 12/04/2021).

World Bank, TerrAfrica and Government of Burundi (2017) Burundi Country environmental analysis: understanding the environment within the dynamics of a complex world – linkages to fragility, conflict, and climate change. https://openknowledge.worldbank.org/handle/10986/28899 (accessed 12/04/2021).

World Weather Online (2021a) Monduli monthly climate averages, Arusha, TZ. https://www.worldweatheronline.com/monduli-weather-averages/arusha/tz.aspx (accessed 12/04/2021).

World Weather Online (2021b) Moshi monthly climate averages, Kilimanjaro, TZ. https://www.worldweatheronline.com/moshi-weather-averages/kilimanjaro/tz.aspx (accessed 12/04/2021).

Wysocki R (2013) *Effective Project Management: Traditional, Adaptive, Extreme*, 7th edn. John Wiley, Indianapolis, IN, USA.

Zarins J (2018) Leading by example – looking to the future for the shelter and settlement sector. In *The State of Humanitarian Shelter and Settlements 2018. Beyond the Better Shed: Prioritizing People*. Global Shelter Cluster, Geneva, Switzerland, pp. 81–87.

Georgia Kremmyda
ISBN 978-0-7277-6468-3
https://doi.org/10.1680/hce.64683.129

Chapter 5
Making humanitarian engineers

Abstract

The chapter summarises the interdisciplinary, interprofessional skills of humanitarian engineers and the role of higher education to provide them by adopting innovative education strategies. The main roles and skills needed in the humanitarian sector are discussed. Shortages of skills in the humanitarian sector refer to lack of technical and soft skills, with problem-solving and analytical skills, leadership, teamwork and interpersonal skills being among the skills most needed. Career pathways to humanitarian engineering may involve roles from governments (e.g. central banks, ministries of finance, rural development and education) to multilateral development institutions (e.g. the World Bank, International Monetary Fund (IMF) and United Nations (UN)), non-governmental organisations (NGOs) and the private sector (e.g. professional services, manufacturing and investment banking).

DEVELOPING SKILLS FOR A HUMANITARIAN CAREER THROUGH SEMI-FORMAL TRAINING FORMATS, USING CASE STUDIES

Douglas White, Hannah Edmond, Carrie Eller and Ellis Lui

Introduction

The humanitarian engineering sector has evolved in the last few decades, with increased expectations and requirements for entry-level candidates. The level of experience and commitment required now to 'break into' the sector can seem convoluted and unattainable for some. This is true in particular for engineers who are currently working in the private sector and looking to support the humanitarian sector, as relevant experience can be hard to attain through the project work and training opportunities typically available in the private sector. Formalised routes such as studying for a master's degree in humanitarian engineering and then deploying directly into the field are also potentially less practical while maintaining an existing career in the private sector.

Humanitarian engineers hold a variety of key roles within the sector (e.g. technical engineer, consultant, project or programme manager), with various input timeframes: full time roles (e.g. long-term deployments), part time roles (e.g. roster arrangements) and remote support roles. These positions can also sit within different technical disciplines, most commonly water, sanitation and hygiene (WaSH), shelter and logistics. Each role and discipline requires its own specific technical competencies and experience. However, there are also more general core skills that are fundamental to working in the humanitarian sector.

Semi-formal training opportunities can provide an alternative, or supplementary, means for prospective humanitarian engineers to develop the required competencies, experience and overall understanding of the humanitarian sector. There are a variety of examples of these opportunities, including NGO–private sector partnerships, mentoring, volunteering, practical simulations and hands-on training. Combinations of these can be used by engineers interested in the humanitarian sector to improve understanding of the context and develop many of the required competencies.

Target competencies and experience

Target competencies

Working in emergency relief requires both professional competence and personal resilience (Davis and Lambert, 2002).

The Core Humanitarian Competency Framework (CHCF) (Consortium of British Humanitarian Agencies, 2010), was developed through consultation with 15 member organisations to identify the competencies deemed to be most critical for the delivery of effective humanitarian work. This was reviewed by the Core Humanitarian Standard (CHS) Alliance in 2017, which determined that the CHCF is 'fit for purpose, adds value and is highly relevant for staff development and humanitarian efforts in general' (CHS Alliance, 2017a). This report also led to the recommendation to promote these competencies as a global standard to be adhered to (CHS Alliance, 2017b). The CHCF outlines six core competency domains

- understanding humanitarian context and applying humanitarian principles and standards
- achieving results
- developing and maintaining collaborative relationships
- operating safely and securely at all times
- managing yourself in a pressured and changing environment
- demonstrating leadership in humanitarian context.

These competency domains are fundamental for anyone working in the humanitarian sector, and are broken down further into outcomes and core behaviours. These core competencies mainly focus on softer skills, highlighting the significance of teamwork, people skills and understanding the context.

The CHCF also recognises the need for these core competencies to be combined with professional and organisational specific competency frameworks aligned with the role being undertaken. For example, the Urban Competency Framework (Global Alliance for Urban Crises, 2019), or the requirement for specific technical and professional experience and qualifications (e.g. Chartered Engineer status).

Through a series of semi-structured interviews with their members, RedR UK has identified the following skills that humanitarian engineering candidates generally need to strengthen to be effective in the field: (a) understanding the humanitarian sector and context; (b) soft skills, including project management, leadership, communication and intercultural relationships; and (c) local and sustainable capacity building (i.e. looking at the full lifecycle of a project).

The review also noted the need for the following support to those working in the field to enable continual development and learning: (*a*) technical skills development, (*b*) technical support, (*c*) mentoring and (*d*) mental health support (RedR UK, 2018).

In *Engineering in Emergencies* (Davis and Lambert, 2002), the authors also emphasise the importance of softer skills and personal resilience as well as technical ability. Here 'personal resilience' is used as a term to reference personal planning, healthcare and stress management, as well as the ability to work in cross-cultural scenarios.

NGOs are increasingly moving to develop local staff, for example in 2018, 84% of Médecins Sans Frontières staff were locally hired field staff, with the remaining 16% comprising international staff split evenly between field and headquarters roles (Médecins Sans Frontières, 2018). As a result, international staff roles in the field are increasingly focused on managing, training and mentoring local staff to develop skills and improve local resilience. Therefore, the importance of, and requirement for, humanitarian engineers to have core competencies, including cross-cultural and team-working skills, is growing.

Reed and Fereday (2019) summarise competencies for humanitarians into three groupings

■ context – humanitarian, development, high-income country
■ sector – specialist and specific (e.g. engineering, health and education)
■ function – generic and cross-cutting (e.g. manager, communication skills and personality traits).

This can be mapped against the CHCF's structure as

■ context – organisation- and location-specific competencies
■ sector – role-specific technical competencies
■ function – core competencies and management competencies.

Both the CHCF and Reed and Fereday's paper are part of a movement to increase professionalisation and accountability within the humanitarian sector, a motive shared with the Sphere standards (Sphere, 2018). The Sphere standards act as a core humanitarian standard, aligning with the Humanitarian Charter (Sphere, 2018) and promoting quality and accountability within the humanitarian sector through outlining a set of minimum standards to be adhered to. Awareness and experience of implementing these core standards is fundamental in ensuring the delivery of quality humanitarian response.

Experience

There is commonly a requirement for previous experience when applying for roles in the field. For example, the majority of intake positions advertised on Relief Web (UN Office for the Coordination of Humanitarian Affairs, 2021) require a minimum of 'two years of relevant professional experience'. This highlights a paradoxical scenario, whereby experience is required to gain experience. There are a number of forms of experience that can be beneficial when applying for roles in the humanitarian context, both for softer skills and technical knowledge. Some examples include

- working overseas
- cross-culturally sensitive work
- working with stakeholders and the public
- conflict resolution
- working in another language
- management
- teamwork
- adaptability and resilience to cope with uncomfortable, stressful and insecure working and living conditions.

Training for personal and professional development

Engineers of all types have to be multi-skilled. To complement technical expertise, an engineer also has to be a good communicator, be aware of the commercial and legal implications of their work, and also be able to manage people and/or resources. An educational qualification goes a long way to meeting these competencies, but there are some skills that cannot be taught in a classroom.

In order to develop this broad range of skills, engineers must undertake training alongside their other career responsibilities. This is often called continuing professional development (CPD). It comprises the acquisition and development of the special skills and approach needed to be effective as a professional (Figure 5.1). Pursuing CPD is therefore a fundamental part of practising as an engineer; in fact, this is generally an explicit requirement for attaining an engineering professional qualification (ICE, 2019).

CPD in the private sector is typically undertaken in parallel with other career responsibilities due to the need to fit this around core commercial work. Although a variety of formalised training routes for prospective humanitarian engineers are available (e.g. degree courses, full-time internships and more lengthy paid courses run by training organisations), these can be

Figure 5.1 Theory of professional development (Reed and Fereday, 2019)

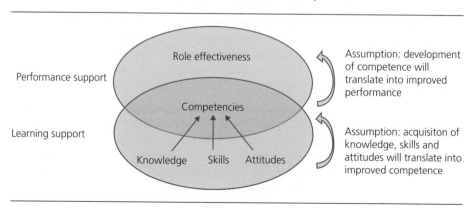

expensive and time consuming – meaning they are not always practical to pursue alongside a private sector career in engineering. This is perhaps why there is a common (mis)conception among prospective humanitarians that the CPD gap between the private and aid sectors is impossibly challenging to bridge.

Function competency grouping

The competencies associated with the humanitarian sector (see the section 'Target competencies and experience' earlier in this sub-chapter) mean that a junior engineer cannot transfer directly into a humanitarian role and expect to be effective. Indeed, even an engineer with extensive professional experience in a non-humanitarian role would not be well equipped to operate in a humanitarian context without a programme of suitable training.

Nevertheless, engineering in the private sector not only provides a basis on which to develop core technical understanding and skills; it is also an opportunity to develop many transferrable 'soft' skills and functions. These include project management; teamwork and leadership; ability to learn on the job; communication; logistics; procurement; stakeholder management; and practical, complex problem-solving.

Sector competency grouping

Candidates are required to develop technical competence and judgement throughout their career. It is essential to build a strong technical foundation before you can be an effective humanitarian, in order to enhance confidence and credibility in the field. This is true even when undertaking a non-technical position due to the broad scope of many humanitarian roles. Engineering in the private sector provides an excellent opportunity to develop technical competence in a controlled and formalised manner, although this may provide less exposure to the resourceful and pragmatic engineering approach generally required in the humanitarian sector.

Context competency grouping

In order to bridge the humanitarian CPD gap, candidates require context-specific training in areas that may not be well served by the typical range of CPD opportunities in the domestic engineering industry.

Semi-formal training has therefore emerged to support early career professionals who are looking to build an understanding of the sector, while continuing to develop their technical skills within the engineering industry. This comprises a variety of learning opportunities including practical (hands-on) training, NGO–private sector partnerships, mentor programmes, conferences and e-learning. Critically, these opportunities showcase the 'thinking on your feet' mentality that is required for humanitarians in the field – with skills such as using minimal resources and being reactive in adverse conditions.

Semi-formal training is defined here as any CPD activity that is not a

- professional/project role that is carried out under the category of 'business as usual' – this would comprise 'for profit' work by the private sector, or the core work of a humanitarian government agency or NGO
- full-time further education (such as a university course).

A list of example formal and semi-formal training is provided in Table 5.1. Typical learning outcomes are provided for each training type, to illustrate how gaps in learning may be addressed. Note that this list of training opportunities is not intended to be comprehensive, and many other opportunities are available to seek out.

Table 5.1 Typical learning outcomes of different training types

Training type	Description	Key benefits
Formal		
University degree course	Master's degree courses are available specialising in humanitarian engineering (University of Warwick, 2021)	Detailed and comprehensive course covering most aspects of humanitarian theory
Internship	Many humanitarian organisations (including government agencies) operate an internship programme. See the section 'Useful resources' for identifying appropriate vacancies	Humanitarian context Working with local staff and organisations
NGO training programme	Formal training programme in advance of deployment. Usually includes security training	Detailed and comprehensive training. However, generally only available to NGO staff
Semi-formal		
Conference	A regular programme of humanitarian conferences is available in most countries, e.g. the biennial Shelter Forum in the UK (Shelter Forum, 2021)	Keeping up to date with innovation and current best practice Building a contact network
Evening learning event (presentation or workshop)	Relevant evening learning events are regularly organised by humanitarian NGOs, including RedR UK (2021c). Many are available to stream online	Keeping up to date with innovation and current best practice
Mentor programmes	RedR UK runs an affiliate programme for aspiring humanitarian candidates (RedR UK, 2021a). The programme matches candidates with an experienced humanitarian mentor, to provide career guidance and training	Humanitarian context Keeping up to date with innovation and current best practice Building a contact network
Hands-on learning	Short practical courses typically run by experienced humanitarians returned from the field	See the section 'Case study: RedR UK Hands-On Weekends'

Table 5.1 (*Continued*)

Training type	Description	Key benefits
Voluntary project work	Small- to medium-scale projects typically with an international development focus. Either run directly by NGOs (e.g. Engineers for Overseas Development (2021), private sector firms or NGO–private sector partnerships (see the section 'Case study: NGO–private sector partnership models')	Humanitarian context Working with local staff and organisations Technical skills development
Field visits	May be required as part of a voluntary project	Local context Adaptability and resilience to new working and living conditions
Technical support	Remote role typically providing support to NGO field operations. See the section 'Mapping a route to the humanitarian sector'	Technical skills development
E-learning modules	Relevant academic modules are available through the e-learning portals of major international universities. Often these modules are available at no cost. Further details are provided in the section 'Useful resources'	Keeping up to date with innovation and current best practice
Reading articles/ publications	Relevant publications are regularly released by humanitarian NGOs, including the Overseas Development Institute (ODI, 2021), Devex (2021), The New Humanitarian (2021) and RedR UK (2021b)	Humanitarian context Keeping up to date with innovation and current best practice

As candidates attend successive humanitarian training events they will expand their contact networks, potentially including established aid workers and/or NGO contacts. These contacts may lead to further opportunities such as mentoring, internships and voluntary support roles.

Keeping up to date with innovation and current best practice in the aid sector is critical if humanitarian candidates are to demonstrate their credibility; therefore, maintaining a baseline of relevant CPD is advisable. Useful resources for this purpose are listed later on.

Hands-on training provides an opportunity to both develop and demonstrate competencies, as well as broaden understanding of the humanitarian context. Volunteering opportunities build on this and include cross-cultural and practical hands-on experience. In the context of semi-formal training, these formats provide an opportunity to simulate the unpredictability of a humanitarian role in a controlled environment. This brings significant value to the candidate in learning how they cope with these situations.

It should be noted that many NGOs have their own bespoke training programmes that are under-taken prior to field mobilisation, which may include detailed security training as relevant for the role.

Finally, exposure to a variety of humanitarian roles through semi-formal training gives candidates a realistic insight into the personal commitments required, and how in future they might balance these with other family and career commitments.

Case study: RedR UK Hands-On Weekends

Background

Since 2013, RedR UK has run a Hands-On Weekend (HOW) in Sussex, UK, alternating each year between Shelter and WaSH themes. The weekends are designed to give participants a practical insight into disaster relief, by taking part in workshops led by expert humanitarians. Additionally, the participants camp outside with limited facilities during the weekend.

The HOW is aimed at beginners (i.e. those who are considering working in disaster relief), and participants have traditionally reflected this demographic. A notable proportion of attendees are private sector junior engineers, with other attendees including nurses, architects, students, international NGO programme officers, teachers, government staff and job seekers. However, over the years the number of experienced international development professionals, attending to learn more about the specifics of Shelter or WaSH in the context of disaster relief, has grown.

The workshops are facilitated by experienced humanitarian practitioners who, together with RedR, develop individual workshop plans for practical hands-on sessions, based on their area of expertise and chosen topics to complement the other sessions. The workshops also incor-porate the latest themes and 'hot topics' in the sector, such as urban disasters, cash and faecal sludge management.

Examples of workshops include constructing Shelter kits, undertaking a solid waste sorting and conveyance task, building a hand-washing station, and surveying and setting out a camp layout. Alongside the technical tasks, various discussions and activities are incorporated to broaden the scope of learning; exercises have included a community engagement roleplay, brainstorming how to adapt latrines for inclusive use, a solid-waste government stakeholder game, discussions on the health impacts of bamboo shelters, and a prioritisation exercise to plan non-food item distribution based on a vulnerability assessment.

In addition to the workshops, talks are given by expert humanitarians on their personal experiences, to demonstrate the reality of working in the sector. Talks have included their specific roles during a particular response, challenges they believe the sector is facing and their experience on cross-cutting themes such as faecal sludge management.

Benefits

A common theme across the workshops is to show participants how complex humanitarian responses are, and that context-specific problem-solving is required to come up with an acceptable solution in a resource-constrained and dynamic environment. For engineers specifically, the HOW shows the difficulty of developing safe, acceptable infrastructure solutions in very challenging situations, and what tools/processes/standards are available to assist (Figure 5.2).

All participants learn detailed practical technical ('sector') skills relevant to the course theme (i.e. Shelter or WaSH). However, the intended areas of CPD are much wider than this – the 'context' learning that is developed by the group discussions that the expert humanitarians facilitate. These discussions help the participants build their understanding of the sector and begin to develop the softer skills that are required for it. Specifically, the HOW touches on context awareness, intercultural working and adaptability/personal effectiveness.

Due to the short duration of the training, these weekends do not fully prepare participants to the level required to be an effective humanitarian. However, they build an understanding of what competencies are required and how they might be applied to scenarios in the field.

Figure 5.2 Participants building a bamboo shelter in the HOW 2017

The HOW is focused on disaster relief, but the skills are generally transferable to the international development sector. In addition, the more generic 'function' skills developed, such as problem-solving, management, communication and teamwork, can be used in many sectors.

The course helps participants to build their network of contacts and understand what potential roles in the sector might match their skill set; this has allowed a number of past attendees to establish their career in disaster relief and international development. It should also be noted that many participants have attended multiple weekends.

Finally, the limited facilities for participants (e.g. the lack of showering facilities) give them a glimpse of the potential reduced comfort they may experience in the field during a response.

In summary, the affordable price and time commitment means participants can 'test the water' of the disaster relief sector and start to develop the skills needed to work in it. The training is designed to be accessible and provide a broad understanding of the sector, with topics specific enough for more experienced participants to gain detailed understanding. The HOW provides a rare opportunity to meet with experts in a relaxed and prolonged 'networking' environment while developing humanitarian competencies.

Case study: NGO–private sector partnership models

Background

The UN estimates that 68% of the world's population will live in urban areas by 2050, compared with 30% in 1950 (UN, 2019). This large shift of people from rural to urban areas will create humanitarian challenges that have not been experienced before. NGOs and humanitarian agencies will need to respond to these challenges swiftly and efficiently. The limited experience of NGOs in urban settings means there is a knowledge gap, which requires external inputs. Partnerships between NGOs and private sector firms are one example of how some aspects of this gap can be bridged.

The complementary strengths of NGOs and the private sector are required for this type of partnership to work, in effect allowing both organisations to be contributors and beneficiaries at the same time. NGOs have pre-existing and established relationships with local partners, allowing them to employ local people and engage with local communities. They are also likely to have an in-depth knowledge and understanding of the cultural and local complexities. Understanding the context is essential to the success of a project, as effective humanitarian work should be driven by the needs of the beneficiaries.

Private sector companies commonly have extensive experience working in urban areas. Alongside a breadth of technical experts, this enables them to provide best practice support to resolve complex challenges, using knowledge of the latest technologies. Additionally, private sector organisations bring experience of contract management, particularly when dealing with liabilities and risks on complex projects, and formalised quality assurance processes. This is important, as humanitarian problems are likely to become more complex, requiring increasing levels of accountability from the sector.

Two examples of where this partnership model has been implemented in the UK are

- Concern Worldwide, an international humanitarian organisation working in both disaster relief and international development, and Mott MacDonald, a global engineering, management and development consultancy.
- Oxfam, a global confederation of 19 charitable organisations, working in both disaster relief and international development, and Arup, a multinational consultancy providing a variety of services, including engineering, architecture, project management and international development.

Both contractual models are based on a modified framework contract – a format that is commonly used across the private sector. The modifications generally regard extent of liabilities, fees and the type of service offered. Both partnerships operate on a not-for-profit basis, unlike most traditional private sector projects.

Benefits

The types of projects undertaken under these models include rainwater harvesting, small dams and clean water gravity systems, and span different countries, including Somalia, Sudan and Bangladesh. Assignment timescales also range from several weeks to greater than 6 months. The diverse range of projects mean there are opportunities for engineers with different interests and expertise to get involved.

Private sector engineers who work on these projects are exposed to challenges outside their usual work, by virtue of the project taking place in developing countries. These same challenges help bring out and develop context-specific competencies required in the field, as outlined in the section 'Target competencies and experience'. From a technical perspective, some of the design tools available for a typical private sector project (i.e. specific design standards, three-dimensional surveying and modelling technologies) may not be relevant in a humanitarian context. These situations encourage first-principles thinking, which is a highly useful skill for humanitarian candidates to develop.

Other benefits to the candidate include

- developing competencies across the 'function', 'sector' and 'context' groupings, including teamwork, cross-cultural communication, societal context and technical appropriateness
- knowledge sharing: a common feature across both partnerships is the importance of having a network of experienced engineers supporting both the junior engineers and NGO staff (including local country staff) – this facilitates knowledge transfer to a new generation of humanitarians and aligns with the wider goal to improve capacity of local staff
- improved career fulfilment and engagement
- overseas experience: where the security situation allows, field visits provide the opportunity to develop competencies related to working overseas, adaptability and resilience.

Other benefits for the private sector include

- staff retention through increased fulfilment and engagement
- business development through demonstrating experience in working successfully in the project region and with NGOs in the sector
- alignment with corporate social responsibility (CSR) objectives.

One potential limitation, however, is access to a network of senior engineers with the relevant experience in the sector. This may not be available in some organisations, and therefore the model appropriateness needs to be carefully considered in the context of each firm. In the two partnerships mentioned, before agreeing to take on each project, the private sector firms undertake their own assessment of staff availability and experience.

In summary, working on a project within an NGO–private sector partnership can be an accessible way for engineers to both contribute to and develop an understanding of engineering in the humanitarian sector, while working in a familiar role and controlled environment. An example of a such as project was delivered by Arup specialists and the local Oxfam staff in the assessment of faecal sludge treatment plants in Cox's Bazaar refugee camp (Bangladesh) (Figure 5.3).

Figure 5.3 Arup specialists working together with local Oxfam staff in the assessment of faecal sludge treatment plants in Cox's Bazaar refugee camp (Bangladesh). (Source: Inigo Ruiz-Apilanez (Arup))

Useful resources

Prospective humanitarian candidates may find the following resources useful to supplement their programme of (formal or semi-formal) training

- RedR UK Affiliate Scheme: RedR UK recognises that it may be difficult for people with little or no experience in the humanitarian experience to gain access to the sector. The programme matches candidates with an experienced humanitarian mentor, to provide career guidance and training (RedR, 2021a).
- RedR UK KnowledgePoint: this online technical support service provides free, expert advice to aid and NGO workers operating across the world. Professionals (of any level of experience) can sign up to offer expert advice on an ad hoc basis (RedR, 2021c).
- Engineers for Overseas Development (EFOD): this charity helps to develop and train junior professionals in the construction industry by assigning them to construction projects in Sub-Saharan Africa (EFOD, 2021).
- Core Humanitarian Competency Framework: the CHCF was developed by the Consortium of British Humanitarian Agencies to identify key competencies to carry out effective humanitarian work.
- Engineers Without Borders UK: this charity aims to inspire, enable and influence present and future engineers to contribute to positive global change, through workshops, design challenges and projects (EWB, 2021).

Online resources include

- DisasterReady.org (http://www.disasterready.org): a free NGO-approved online training platform for humanitarian workers and volunteers.
- Coursera (http://www.coursera.org): free online modules delivered by universities across the globe. Many modules directly relate to humanitarian engineering topics.
- ReliefWeb (http://www.reliefweb.int): a UN website providing information to humanitarian relief organisations.
- Global Peace Careers (https://globalpeacecareers.com): a website dedicated to career-related information, sharing job opportunities at various levels in sectors such as humanitarian aid, international development and diplomacy.
- Voluntary Service Overseas (https://www.vsointernational.org): a not-for-profit international development organisation offering volunteer programmes to deliver development impact.

This is by no means an exhaustive list, and candidates should note that many of these resources are UK based.

Mapping a route to the humanitarian sector

Full-time roles

A great variety of full-time humanitarian roles are available to candidates with the relevant competencies and experience (see the section 'Target competencies and experience' earlier in this sub-chapter). Roles are available in NGOs, the private sector or working directly for national governments. Many roles are not located in the field, but are instead support roles based in the organisation's headquarters or with the funding agency. The contractual nature and length can also vary – contracts can last for the duration of a specific assignment/project, or continue across a number of deployments/headquarter roles. Voluntary positions and internships are also common alongside paid roles. The section 'Useful resources' provides various suggestions for seeking out and applying for humanitarian positions.

Due to the security risks and expectations around effective performance, it is unusual for junior engineers to be offered a full-time humanitarian position in the field without previous relevant voluntary experience abroad and/or training related to the developing world. Vita Sanderson worked as a graduate structural engineer for a private sector engineering consultancy in the UK. During a RedR UK HOW event, Vita was recommended to apply for a voluntary role in Bangladesh. The role was with the NGO UN Accord, which was deploying structural engineers to carry out a number of structural assessments on public and commercial buildings following the 2013 earthquake. Vita took 2 months unpaid leave from her UK position and joined the NGO mission, where she was able to demonstrate the transferrable technical and managerial skills from her career while gaining valuable field experience.

On her return to the UK, Vita continued to gain humanitarian exposure through organising a programme of training events for junior engineers, as well as participating in a (voluntary) project to interconnect private sector engineering firms in order to create a 'humanitarian hub' for support to NGO operations in urban environments. Through these opportunities, she was able to build her network of contacts. After a further voluntary role working under the UN Children's Fund to build schools in Nepal, Vita was offered a full-time paid role by Médecins Sans Frontières, first in Ukraine for 7 months and latterly in South Sudan for 11 months.

The role in South Sudan involved expansion of an existing hospital, in order to provide basic healthcare to the population in the autonomous region of Abyei. Vita was responsible for project coordination, managing a team of expatriates and local staff. This meant that teamworking, training and logistics skills proved invaluable alongside her core technical competency. The remote nature of the role also forced the team to 'think on their feet' and develop a pragmatic approach to overcoming the various engineering challenges.

The transport networks and security situation in South Sudan meant travel to and from the hospital was relatively difficult and, consequently, breaks during the role were few and far between. Vita highlights that career planning is important in the humanitarian sector in order to mitigate the intense nature of many field roles and avoid 'burn out'

> When going into a full-time field role, it's important to have an exit strategy and timescale in mind. This can potentially comprise changing role within the NGO, for example rotating to take on a headquarters position for a few months. Equally, having gained humanitarian field experience is good for your CV, so candidates are in a strong position to step into a private sector role upon their return home, while potentially maintaining a part-time role or providing ongoing ad hoc support [see below]. Try and think a few steps ahead; but be agile enough to change plans if the opportunities present themselves.

Part-time roles

The intense nature of disaster response work can mean that full-time humanitarian roles may not be practical for many technical professionals, especially given the need to balance a career with personal/family commitments. To facilitate shorter inputs, some humanitarian response NGOs maintain a register (or roster) of technical professionals who are trained and ready to deploy at short notice following an emergency. Such registers are often focused around a specific humanitarian discipline, such as WaSH or Shelter.

Such NGO frameworks are set up to allow all parties to benefit from the arrangement

- Private sector firms are able to publicly promote the humanitarian work their staff have undertaken as CSR. Furthermore, they are able to benefit from the professional experience gained by employees mobilised in the field, and are better able to retain staff who feel fulfilled in their diverse role.
- Engineers are able to build their humanitarian skills and CV, and achieve career fulfilment, while maintaining a balance with their non-humanitarian career and personal life.
- NGOs benefit from access to a large and diverse pool of skilled engineers at a relatively low overhead cost (comprising periodic training and human resources inputs). The engineers can be mobilised at short notice and are up to date with current best practice as a result of the periodic training and domestic professional roles.

Joel Westberg works as a senior process engineer for a private sector engineering consultancy in the UK. In addition to his domestic role, Joel is a member of the Red Cross Emergency Response Unit (ERU), specifically within the Mass Sanitation Module specialising in implementing water and sanitation solutions. The role comprises short-term deployments in response to an emergency request from the International Federation of Red Cross and Red Crescent Societies, typically for a period of 4–5 weeks. In 2018, Joel was deployed to Cox's Bazaar in Bangladesh, to provide emergency faecal sludge management to refugees from Myanmar.

> There was no textbook or formal guidance available that was relevant to what we found on the ground in Bangladesh; we had to improvise based on our previous experience. I found all the transferable skills that I had acquired through both my professional career and the various semi-formal trainings proved invaluable here.

Joel indicates to the Red Cross when he available to deploy over the coming 12-month period, in conjunction with his domestic employer. This agreement is then written into his domestic contract in order that that he can be released for short periods to support the ERU, during which he is paid directly by the Red Cross. Joel considers that this arrangement provides him with a sustainable balance between his career objectives and personal commitments.

In addition to the irregular (reactive) requests for deployment into the field, the Red Cross provides a preparatory formal training course to all unit members. It also provides periodic 'refresher' training that allows unit members who have not recently deployed to stay in touch with their humanitarian network and maintain current best practice.

Remote/ad hoc support roles

Remote support roles are also available for candidates who are

- looking to build their humanitarian experience (through semi-formal training)
- between field roles and looking to maintain their competencies
- motivated to contribute to the sector but are unable to take up a full-time or part-time role.

A variety of remote support roles are available to candidates, such as NGO–private partnerships and technical support opportunities. In particular, open-access remote support portals are available through NGOs to support field operations. RedR UK operates its KnowledgePoint service to link requests for technical support from humanitarians with a network of technical professionals around the world. There is no minimum experience requirement to contribute to this service.

Bridging the career gap

As discussed earlier, there is a common perception among private sector engineers that the experience gap between the humanitarian and aid sectors is impossibly challenging to bridge. However, the attributes and skills acquired in the course of a domestic engineering career are transferrable and relevant; the gap can therefore be closed with supplementary training (either formal or semi-formal). Joel feels that the gap can sometimes be a misconception

> Prior to applying for the Red Cross ERU role, I didn't have any formal disaster relief experience in the field. But I was able to demonstrate that I had pursued many opportunities that had provided a tangible humanitarian experience, which I was able to talk about in the interview. These included short-term internships and voluntary roles with NGOs at both university and graduate level. I was also able to demonstrate my transferrable skills acquired during my work in the private sector; which I only understood once I'd had sufficient [informal] exposure to how the humanitarian sector operates.

Vita highlights the importance of gaining first-hand experience of working in a humanitarian context, including voluntary roles in an international development (rather than emergency response) context

> Field experience is a must. Often the most difficult people to work with in the humanitarian sector were those who had never been in the field themselves.

Vita also notes the changing focus of the sector to engage and train more local staff

> There is a general recognition that the 'fly in, fly out' aid model is not financially, environmentally or socially sustainable. Therefore, the emphasis is changing from direct technical intervention to training and capacity building of local teams. More than ever, it is important to demonstrate competence in training and mentoring. However, often candidates are experienced in training people from exactly same background as themselves; but be mindful that this isn't always the case in a humanitarian context with the additional language and cultural barriers to overcome.

Both Joel and Vita agree that building a network of contacts should be a key focus in parallel with developing humanitarian competence and experience, as the people met during training events and conferences are often the best route to accessing future opportunities as they arise.

Their advice to junior technical professionals was not to worry about becoming a full-time humanitarian too early in the course of a career. In the early career phase, a lot of the benefit gained through such roles is primarily for the candidate themselves – through strengthening of competencies (not just technical) and building experience. The target groups (affected persons) may receive little actual benefit from a junior professional, as it takes time to develop technical competence and judgement; it is therefore useful to build a strong technical foundation before you can be an effective humanitarian. Time spent working in the private sector can help achieve this, which – alongside a programme of semi-formal training to develop humanitarian competencies – will allow a more sustained impact in the longer term.

Conclusions

There are many routes for engineers seeking to work in the humanitarian sector. However, with increasing accountability and professionalisation, acquiring key competencies is fundamental to enabling effective delivery of projects. As a result, competency frameworks such as the CHCF (Consortium of British Humanitarian Agencies, 2010) have been developed to provide guidance in these requirements. Competencies include technical role-specific abilities and soft skills such as management, communication and personal effectiveness.

For many in the private engineering sector the perceived gap between the sectors can be daunting. However, semi-formal training opportunities can enable prospective engineers to develop an understanding of the humanitarian sector, broaden their network of contacts and build the required competencies. This can assist with moving into the field while continuing to acquire technical knowledge and experience as well as transferrable softer skills in their existing job roles.

There is no single prescribed formula for success. There is also no requirement to rush into a full-time humanitarian position very early in the course of a career. It takes time to develop technical competence and judgement; it is therefore useful to build a strong technical foundation before becoming an effective humanitarian. Candidates should adopt a sustained approach to achieving their goal, and make use of opportunities that match their competence and experience levels – giving them the best chance of succeeding in a rewarding and long-lasting humanitarian career.

HUMANITARIAN ENGINEERING IN INDIA: STATUS AND PROGNOSIS

Seema Singh

Introduction

Technology has received a lot of attention in the capitalist framework, as it reduces labour power and generates more profits, which are ultimately accumulated by capitalists. It also gives capitalists more control over the system (Gera and Singh, 2019a). Across various schools of economic thought, technology has been recognised as the means to increase productivity for all (Gera and Singh, 2019b). In other words, one of the influencing factors for the growth of an economy is the availability of technology, which, to a certain extent, depends on the status of engineering education. Engineering education enables human resources to innovate, operate and maintain technology. In short, engineering education has a key role in the development paradigm of a country.

Technology, although considered neutral in its approach, has had and continues to have an impact on society and production. The use of technology for large-scale production, commonly known as the first industrial revolution, has had a toll on natural resources as not enough time was given for regeneration due to excessive use. The environment equilibrium has been disturbed, leading to environmental degradation. However, the traditional development framework never saw this as a problem but as by-product of development. It was considered to be a cost that society had to pay for development. the engineer Herbert Hoover, the 31st president of the USA, was among the first leading figures to express concerns over environmental degradation (Singh, 2014).

Even though the laws and principles of natural science are place objective – Newton's law is the same throughout the globe – when applied to develop a technology, these become place subjective when using it: the same technology is not appropriate in all places across the globe. Another aspect of technology to be considered seriously is its contribution to sustainability in relation to economic, social and environmental aspects. The impact and measurement of sustainability has a different timeframe; economic sustainability can be immediately calculated, but it takes a comparatively longer timeframe for social sustainability to be measured. If a technology is not sustainable – either economically or socially – it will not be produced or used. However, the case is not the same for environmental sustainability. It takes a much longer time to become visible and have a measurable impact. Very often, producers decrease cost by making an environmentally non-sustainable product or service, and bargain economic sustainability for environmental sustainability.

During the pre-globalisation period, countries, including India, imposed restrictions on technology entering the domestic market. But, in the era of post-globalisation, when information and communications technology (ICT) became pervasive in the production process, geographical boundaries have lost significance. Even economic boundaries have blurred. In such circumstances, the pressure of fast economic development at the macro level and sustaining the market at the micro level have resulted into two changes. First, the enhanced use of technology by domestic players and, second, many multinational and transnational enterprises have established a base in India. Alongside, ICT has allowed the producers of the global north to shift the labour-intensive segment of their production processes to developing economies, where these activities can be done at a lower cost and without compromising quality. India has an advantage in this regard owing to its large English-speaking population (Singh, 2005).

One outcome of this transformation is the rapid urbanisation and growth of unplanned peri-urban areas. At present, about 55% of the world's population is living in urban areas, and this is expected to grow to 68% by 2050. The world's population will be highly concentrated in few countries, with India leading the list – 416 million is projected for the India urban population (UN, 2019). Naturally, the growth of the urban population will not have a uniform background and equitable access to resources. In fact, this growth will include significant rural–urban migration or migration from smaller to bigger cities, from the lower segments of the society in search of opportunities.

As per the Oxfam India inequality report (Oxfam India, 2018), income disparity is increasing rapidly in India. India's top 1% of the population owns 73% its wealth, and, on the other hand, 70.6 million people live in extreme poverty – on less than $1.90 a day. Technological progress and the resulting rise in the skills premium (positives for growth and productivity) and the decline of labour market institutions have contributed to this inequality. However, earlier IMF work has shown that income inequality matters for growth and sustainability (Dabla *et.al.*, 2015). Even the 17 UN Sustainable Development Goals (SDGs) (UN, 2021) may be segregated into three groups: economic, social and environmental (Table 5.2). Nevertheless, many of the SDGs are cross-sectional within any two or all three pillars of sustainability.

Goals 1 and 10 belong to the economic sustainability pillar, but achieving them is important to attain social sustainability too. No society can remain sustainable with large income inequality. In fact, this has been realised by the Government of India, and the 11th Five Year Plan of India exclusively discusses inclusive growth. Since independence, the Government of India has emphasised the importance of domestic technology, and a network of science and technology institutions was established for rigorous scientific research. For the time being, the Government of India has decided to depend on borrowed technologies from European countries. These countries have less labour and more capital while in India the conditions are just the opposite. This has resulted in employment for a highly skilled workforce, with significant unemployment among the less skilled or unskilled. This underprivileged and less educated

Table 5.2 Segregated UN SDGs

Economic sustainability	Social sustainability	Environmental sustainability
Goal 1: No poverty	Goal 3: Good health and well-being	Goal 6: Clean water and sanitation
Goal 2: Zero hunger	Goal 5: Gender equality	Goal 7: Affordable and clean energy
Goal 4: Quality education	Goal 16: Peace, justice and strong institutions	Goal 11: Sustainable cities and communities
Goal 8: Decent work and economic growth	Goal 17: Partnerships for the goals	Goal 13: Climate action
Goal 9: Industry, innovation and infrastructure		Goal 14: Life below water
Goal 10: Reduced inequalities		Goal 15: Life on land
Goal 12: Responsible consumption and production		

workforce cannot afford to be unemployed, and take up any job they can get – but returns from their work do not provide sufficient income for healthy living. The corporate sector, supported by qualified professionals, takes decisions about technological upgrades, but small and lesser enterprises rarely know what type of technology will advance their productivity. There is a need for social innovation to be inculcated in engineering higher education institutions (Singh, 2013). However, social aspects of innovation and technology are disconnected in engineering education, and social sciences/humanities education are decoupled as subjects. There is a need to assimilate such subjects for social change.

At least ten out of the 17 UN SDGs need the direct involvement of engineers for their success (Table 5.3). So, engineers across all branches are at the core of the development strategy in the present scenario. As an example, to achieve self-sufficiency in agriculture, India needs improvements in supply chain management and storage. Mechanical and electronic engineers alongside agriculture engineers, bio-technology and genetic engineers need to deliver combinational and innovative outcomes. SDG 3 relates to good health, but interconnects strongly with SDG 2. Along with medical scientists, electronic engineers are expected to work towards achieving the goal.

Table 5.3 UN SDGs and branches of engineering

SN	SDG	Description	Related branch of engineering
1	Goal 2	Zero hunger	Agricultural engineering
			Biotechnology
			Genetic engineering
2	Goal 3	Good health and well-being	Biomedical engineering
			Genetic engineering
			Biotechnology
3	Goal 6	Clean water and sanitation	Mechanical engineering
			Architecture
			Civil engineering
4	Goal 7	Affordable and clean energy	Electrical engineering
			Electronic engineering
			Information technology
5	Goal 9	Industry, innovation and infrastructure	Mechanical engineering
			Civil engineering
			Production engineering
			For innovation: all branches of engineering

Table 5.3 (*Continued*)

SN	SDG	Description	Related branch of engineering
6	Goal 11	Sustainable cities and communities	Information technology
			Computer and software engineering
			Architecture
			Civil and environmental engineering
7	Goal 12	Responsible consumption and production	Production engineering
			Mechanical engineering
			Mechatronics engineering
			Nanotechnology
8	Goal 13	Climate action	Environmental engineering
			Civil engineering
9	Goal 14	Life below water	Biotechnology
			Genetic engineering
10	Goal 15	Life on land	Biotechnology
			Genetic engineering

Source: Singh (2019).

Status of engineering education in India

General

Engineering education plays a critical role in the economic and social development of a country. At the time of Indian independence in 1947, facilities for undergraduate courses in engineering were available mostly in the disciplines of civil, mechanical, electrical, chemical, metallurgy, mining and telecommunications engineering. The number of institutions offering such courses was only 44 (Ramachandran and Kumar, 2003) (Table 5.4).

Growth of engineering education during the pre-globalisation period (before 1991)

After independence, as per the recommendation of the Sarkar Committee (1945), a planned effort was made for the growth of the higher engineering education system to meet the requirements of post-war development. Within a brief period of 3 years (1947–1950), nine engineering colleges were founded, which rose to 102 within the next 10 years. Engineering education was also seen as a source of innovation. The Thacker Committee (1959–1961) recommended the establishment of Indian Institutes of Technology. From the 1960s onwards,

Table 5.4 Status of degree level engineering education in India

Up to the year	Number of institutions	Intake
1947	44	2 500
1950	53	3 700
1960	102	16 000
1970	145	18 200
1980	158	28 500
1990	302	66 600
2000	880	2 28 511
2010	2 872	1 071 896
2020	3 048	1 329 159

Sources: up to 2000, Ramachandran and Kumar (2003); for 2010, Indian MHRD (2010); for 2020: All India Council for Technical Education (2021).

the focus shifted to research through postgraduate education and PhD courses. (Shaha and Ghosh, 2012). During 1950–1960, the number of engineering colleges offering bachelor of engineering or bachelor of technology programmes rose to 102, which further increased to 145 by 1970, 158 by 1980 and 302 by 1990 (see Table 5.4). In this period the government was the major player in the development of tertiary education.

Growth of engineering education during the post-globalisation period (after 1991)

After globalisation, with the opening up of the Indian economy and the use of ICT, the demand for engineers increased manyfold, as many multinational and translational companies established their production units in the country. There is also the advantage of a huge market due to the large population size. ICT has facilitated transfer of some production processes from the global north to India, where these can be done to the same quality but at lower cost – commonly called business process outsourcing. The education system has responded accordingly. By the year 2000, the total number of institutions has reached 880. Most of these later institutions came through private initiatives (Ramchandran and Kumar, 2003). By the end of 2010, the number of institutions had risen to 2872 and by 2020 this had increased to 3048 (see Table 5.4).

Humanitarian engineering and its status in India

General

Humanitarian engineering may be defined as the integration of engineering with the social sciences and humanities through tools, techniques and appropriate pedagogy to equip engineers to face humanitarian challenges – social and economic as well as environmental – through engineering interventions to promote the well-being of those facing such challenges. The need for the integration of engineering with the humanities, social sciences and management has started to become recognised throughout the world, including India.

Languages and social sciences in engineering education

With the above in mind, the All India Council for Technical Education (AICTE) set down norms for the inclusion of social science education. This body stipulated that engineering-related programmes should include 5–10% of the total course content, and it would be desirable to have a minimum of one course in each of the areas below (Singh, 2005)

- languages/communication skills
- humanities and social sciences
- engineering and principles of management
- National Service Scheme (NSS), National Credit Core (NCC) and rural development.

Such courses should cover at least the basics. Advanced courses, if offered, may be from the time allotted for core engineering subjects. However, a study among engineering colleges in Delhi and Haryana, conducted under the auspices of the National Technical Manpower Information System found that no institutions provided a humanities and social sciences content of more than 10% of the total course content, but better ranked colleges had more options. English was being taught in all institutions in the first semester of almost all branches of engineering. Economics, psychology, sociology and other topics in the social sciences were introduced in the second or third year of engineering courses, and some of them included more than one of these areas. Within a college, different branches of engineering may provide different subjects in the social sciences (Singh, 2002).

A deeper analysis revealed that often such subjects have been included simply for the sake of inclusion, and little thought had gone into the selection of the social science topics. They were fragmented, and sometimes seemed of limited relevance to engineers. The fragmented nature of these topics, a lack of structured format and dissimilarities between subjects taught by the engineering educational institutions (even in the same city, e.g. New Delhi) led to the unavailability of suitable textbooks. This was also clear from the list of suggested textbooks in the printed syllabus of many colleges, which typically included only those books written for students on degree courses in the social sciences, such as economics (Singh, 2002). Although geography and psychology are both very important, they were not being taught anywhere. Geography has a very special status in deciding inputs in manufacturing. Its significance has increased in the era of globalisation. International competition has compelled firms to think globally and work locally. A new term, 'glocal', has been coined. In the present economic environment, cost effectiveness and competitiveness can be achieved only if process and product can be developed by using locally available resources. There is also a need for psychological understanding by engineers. Working in a project requires team effort. Engineers can only function with the help of a group of people, including their subordinates, peer group and superiors, if they are to provide an effective service to their clients, or to groups within society. They need to have a psychological understanding of group dynamics (Singh, 2005).

Environmental science

Continuing problems of pollution and degradation of the environment, including specific issues such as economic productivity, national security, solid waste disposal, global warming, the depletion of the ozone layer and loss of biodiversity, have made everyone aware of environmental issues. The UN Conference on Environment and Development held in Rio de Janeiro in 1992 and the World Summit on Sustainable Development in Johannesburg in 2002 drew the attention of people around the globe to the deteriorating condition of our environment. No citizen of Earth can afford to be ignorant of environment issues. Environmental management has captured the attention of health care managers. Managing environmental hazards has become increasingly important. Recognising this, the Supreme Court of India directed the University Grants Commission (UGC) to introduce a basic course on the environment for every student at undergraduate level for all courses and faculties in India. Accordingly, the matter was considered by the UGC, and a 6-month compulsory core module course in environmental studies was prepared (UGC, 2003) and implemented in all faculties of universities/colleges in India from the academic year 2007–2008.

Globalisation and the opening of economies along with the application of ICT in business processes have made markets very competitive. The corporate sector and multinational companies are continuously upgrading technology as their survival strategy. Ultimately, they have raised the bar for technology adoption. Innovation has become the buzzword for remaining competitive, which creates a lateral pressure on existing micro and small enterprises. Such enterprises may not be know what type of technology they require, and even if they are aware, they will not be able to purchase it at the market price. A need for the availability of technology at low cost is apparent.

However, the prevailing system has been mainly developed to cater for the need of big businesses and corporate entities, neglecting those in lower segments of society. The poor are caught in a vicious circle of poverty and the need for engineering intervention to enhance their productivity. On the basis of existing programmes initiated by various stakeholders in India and the experience of many large-scale projects that have resulted in huge displacements of people, this sub-chapter will next discuss how students may be sensitised through humanitarian engineering courses towards those marginalised and in making projects that are more inclusive and humane (Singh, 2013).

Choice-Based Credit System

UGC initiated its Choice-Based Credit System (CBCS) for all undergraduate and postgraduate level degree, diploma and certificate programmes, including engineering studies, in 2015–2016. Apart from the core courses, which are compulsory, students take elective courses, called open electives, from a pool of subjects (Indian MHRD, 2014), which

- are supportive of the area of study
- provide an expanding scope
- offer an exposure to another discipline/domain
- enhance students' soft skills.

These provisions give students the opportunity to study subjects in other domains that may not have a direct bearing on their humanitarian considerations, but will broaden their mindset and perspective of looking at things. However, studying stand-alone subjects, even though they provide wider knowledge, prevents their integration into the core engineering subjects.

Integration of humanitarian aspects in engineering education

General

Realising the need to familiarise students in India with real-world problems, the Kothari Commission (1964–1966) recommended that universities (higher education institutions today) (Kothari Commission, 1966)

- seek and cultivate new knowledge, to engage vigorously and fearlessly in the pursuit of truth, and to interpret old knowledge and beliefs in the light of new needs and discoveries
- provide the right kind of leadership in all walks of life, to identify gifted youth and help them develop their potential to the full by cultivating physical fitness, developing the powers of the mind and cultivating the right interests, attitudes, and moral and intellectual values
- provide society with competent trained men and women who will also be cultivated individuals, by imparting a sense of social purpose among students along with training them in agriculture, arts, medicine, science and technology and various other professions, in order to promote quality and social justice, and to reduce social and cultural differences.

During the birth centenary year of Gandhiji, the father of the nation, in 1969, the Ministry of Youth Affairs and Sports launched its National Service Scheme (NSS) across 37 universities with 40 000 students. This rose to 3.2 million students from almost 300 universities. The NSS is a Central Sector Scheme of the Government of India, Ministry of Youth Affairs and Sports. It provides opportunities to students in the 11th and 12th grades in schools at +2 board level and to students at technical institutions and graduate and post-graduate students at colleges and universities in India to take part in various government-led community service activities and programmes. The sole aim of the NSS is to provide hands-on experience to young people in delivering community service (Government of India, 2021). The prime focus of the programme is to broaden the mindset of students through participation in community service. However, participation is optional, and the nature of the work done and the approach are very casual. Students can participate in activities related to rural development (Singh, 2016).

More recently, the National Assessment and Accreditation Council of India started consider the level of interaction between universities and communities at the time of granting accreditation to the universities and institutions. The level of involvement with the community through research and other student activities contributes to a deeper understanding of a number of issues, from income inequality to gender disparities and their context, and makes students better equipped to take up real-world problems. Students are exposed to sustainable practices from the community, and the community benefits from the knowledge and enthusiasm of students and of the institution, which ultimately leads to a win-to-win situation, with successful outcomes in terms of generating knowledge useful for the students as well as for the community (NAAC, 2019). There are many national and international students' organisations that do similar work. However, there are some concerning issues related to student participation. First, students prefer to participate when they are free during the academic calendar, but the community may need them when they are unavailable. So, their involvement may not yield its best. Second, and very important, is funding. Singh (2016) has discussed that only sponsored programmes are able to continue long term, as unsponsored initiatives cannot be sustained (Singh, 2016).

The Ministry of Human Resource Development (MHRD) has initiated its flagship programme Unnat Bharat Abhiyan (UBA 2.0) on 11 November 2014, in which universities collaborate with rural administrations to deliver activities to further rural development. The programme has proved to be very successful. Altogether, 98 technical and 45 non-technical institutions participated in the programme, with the involvement of 715 villages from 29 states. Fifteen mentoring institutions were identified. Because of this success, UBA 2.0 was relaunched on 25 April 2018. At present, 428 technical and 260 non-technical institutions are participating from 33 states, where a total of 3440 village have been covered (Indian MHRD, 2020).

Corporate social responsibility

In 2014, the government of India modified its company law to make CSR activity mandatory for high-earning companies. Every company irrespective of ownership has to spend 2% of its average net profit for the immediately preceding three financial years on CSR. This development work has a net worth of Rs 5 billion or a turnover of Rs 10 billion or a net profit of Rs 50 million (Singh, 2016). Section 467 of company law provides details of CSR activities, which includes support for technology development in the academic institutions and university. Research on university social responsibility has identified lack of funding as being the biggest barrier to enhancing and improving the quality of activities taken up by students (Singh, 2013).

Humanitarian challenges and engineering education

At the time of writing (2021), climate change, a global pandemic (COVID-19) and emerging challenges of the fourth industrial revolution are some of the issues with a direct impact on communities. Engineering education is being called on to prepare students for tackling such challenges. Apex regulatory bodies for science and technology are attempting to increase the exposure of interdisciplinary, interprofessional subjects to engineering students. The AICTE has allowed students to take up courses through MOOCs – massive open online courses (Indian MHRD, 2015). The Department of Science and Technology offers several programmes within a science and technology framework borrowed from socio-economic programmes (Indian Department of Science and Technology, 2020). Student and faculty exchange programmes are allowed.

With a vision to expand the use of artificial intelligence (AI), NITI Aayog (a policy think tank of the Government of India) organises hackathons to source sustainable, innovative and technologically enabled solutions to address various challenges in solving current issues. To take the initiative forward, NITI Aayog partnered with Perlin (a Singapore-based AI start-up) to launch the AI 4 All Global Hackathon, and invited developers, students, start-ups and companies to develop AI applications to make significant positive social and economic impact for India (NITI Aayog, 2018). As an example of social and ethical considerations in scientific and technological progress, UNESCO (2018) recognises that AI has become a major research area in science and technology, but without its ethical aspects being explicitly discussed. Brundage (2016) has stated that the societal dimensions of AI can be enriched within a concept of 'responsible innovation,' and discusses not only the technical domains (e.g. nanotechnology, synthetic biology and geoengineering) but also the social context and consequences of AI application. Engineers and scientists need to systematically be mindful of how advances in technology affect the public, and must be aware of future opportunities for research and practice across societal dimensions (Brundage, 2016).

Conclusions and recommendations

In general parlance, engineering is the application of the laws and principles of the natural sciences to improving the quality of life of human beings. So, the humanities and social science should always be part of engineering education programmes. Nevertheless, research across a number of engineering and technological institutions in India have shown that such subjects are included merely for the sake of inclusion, without giving much thought to integrating them efficiently to curricula.

Engineers arrived at the centre of the development paradigm at the advent of globalisation. At least ten out of the 17 UN SDGs need direct involvement of engineers, if they are to be achieved by 2030. These goals, as well as similar targets, norms at the global level and their technological solutions have increased the need for a deeper understanding of the importance of sustainability – environmental, social and environmental aspects in engineering solutions. Few bespoke humanitarian engineering programmes exist, providing interdisciplinary curricula from across engineering, sciences, humanities, social sciences and management (University of Warwick, 2021). In those that do exist, the subjects are not only taught as stand-alone subjects but are fully integrated with each other and applied to real-world problems. The main objective of such programmes is to capture indigenous knowledge and practices.

DESIGN THINKING FOR SOCIAL IMPACT

Robert O'Toole

Introduction

Melbourne, October 2018. We are in a vast lecture theatre at Monash University, awaiting the start of a session as part of the Monash Education Academy's annual conference. But this is no ordinary conference session, and no ordinary lecture theatre. The participants are arranged in groups of six, sitting around white tables oriented towards the usual large projection screen at the front of the room. We can all easily look to the front of the room, but the arrangement of chairs pulls us towards collaboration within our groups. The tables are 'writable' – that is to say, wipeable whiteboard surfaces, with accompanying pens, and a supply of stationery, giving an immediate nudge towards collaboration and contribution. We're not here to just listen, we are here to take part, to contribute together. So this is not really a lecture theatre at all. It is a vast team-working space facilitated by the shared focal-point at the front of the room, and a brilliant facilitator. And this is not really lecture, it is design thinking for social impact, in action.

The session begins with the presenter introducing herself as Ana Stankovic, from the Monash University IT department, eSolutions. The aim is to collectively rethink the Monash student experience so as to fit better with the needs of the highly distributed and increasingly diverse community that it serves. The emphasis is on designing education for positive social change, with a realistic appreciation of the world as it is today. Australian universities operate within a socio-economic environment with many challenges. During the conference we talked about the urgent need to support Aboriginal communities, as well as migrants and other under-represented groups. This is not so much about bringing people into the university, as taking the university out into the world with new connections that are sustained and transformative. That necessarily includes technology and the engineering perspective. But it is most definitely a 'human-first' agenda.

I am here in Australia with Graeme Knowles from WMG (a manufacturing and design-oriented department of the University of Warwick), as a collaboration with Irwyn Shepherd of Monash University (coordinating its virtual-reality (VR) facilities and projects), developing a framework for the use of VR in engineering education. Our team is multidisciplinary, including engineering, education, design research, computing, psychology and philosophy. Our aim is to use extended-reality techniques (which cover the full spectrum of virtual and augmented immersion) to transform the student experience. It should be possible to use VR, for example, to allow students to experiment with otherwise inaccessible or even impossible engineering scenarios. One example, identified through our workshops at Monash University, would be to give students a realistic experience of what it is like to feel and hear a railway bridge in use for the first time – the nerve-jangling experience of a vast assemblage of materials bedding in as weight and movement are applied. In reality this is a rare opportunity. The virtual experience could be a way of engendering 'wonder in engineering', to address a common problem in the curriculum, that students can become disconnected from what makes engineering so exciting. Graeme likes to put this in the context of a 'head, hand, heart' approach to engineering. We need to ensure that students have their heart in the subject. But it is also a way of widening access to real engineering activities, to give every student an opportunity to understand the impact of design decisions (the 'head' part of the formula). If we make the VR experience interactive, then perhaps it can also engage the 'hand' of the engineer. The visualisation team at WMG has already developed remarkably convincing prototypes that allow students to build motors in VR. We can use this, for example, to allow them to experience variations in the process that illustrate 'lean' manufacturing principles. Ultimately, these developments could bring more people into the engineering fold. That could include community groups that would otherwise simply be on the receiving end of engineering projects. We are designing for positive social engagement.

Returning to the session in Melbourne, the presenter begins with an understated announcement that we are going to use a design thinking approach to work together on the challenge of rethinking education. We have used design thinking methods already for our project. But not on the scale of this session. Personally, I have a lot of experience with this approach, including writing a PhD thesis titled *Design Thinking for the [Re]making of Higher Education* (O'Toole, 2015). I can see that the Monash team knows what it is doing, and that it is not unusual for it to be working in this way. Most impressively, there is an interactive dynamism to its approach that immediately draws us all into a kind of collectivity. The presenter is very much a facilitator. We are immediately presented with a shared 'design canvas' upon which our ideas will be developed. Instead of a conventional lecture theatre lectern, this room has a large round table at the front. Again, it is writable. When the facilitator writes on the table, it is projected through an overhead camera onto the big screen so that we can all see the emerging model. The table is also equipped with sets of charged and connected iPads that may be distributed around the room. In this case, the facilitator is 'working the room' with a radio mic. She is easily able to walk between tables, hand over the mic, and get participants to speak. This seems effortless. And the flow of conversation seems just that – conversation, rather than nervously addressing a lecture theatre. She can also easily show sticky notes and diagrams from each of the tables to the whole audience. Her technique in this case is to take sticky notes from each table and add them to the emerging assemblage on the table at the front, with annotations and connections hand drawn. This is done with an engaging dynamism that is both exciting and establishes trust in the process. In a remarkably short time we have developed something together on the screen. But, more importantly, we've got into a mindset that gets us thinking about the detail of what might be possible in relation to the challenges we are considering. We are talking about detail, about things that matter to real people. We are sharing stories and empathising with real people facing real-world problems. But we are not getting bogged down in the detail or the extent of the challenges. Creativity is happening. There's a sense of progress. Not that anyone thinks we will create solutions now. But rather, that we are moving in the right direction, building shared design knowledge, and contributing to a movement in the right direction. Ana is working the room brilliantly. As a result, it feels as if the event is very much fluid and emergent, but there is structure to what we are doing, typical of design thinking.

Experiencing design thinking

If ever I'm asked to define design thinking, which should be easy for me to do as I practise it and teach it, I say 'don't define it, experience it'. Hopefully this short report on a good example of design thinking in action conveys enough of the experience to illustrate *some* of its key features. Facilitation, visualisation, the physical organisation of participants, are used to get people focused on real-world challenges that are iteratively redefined as we explore them. In this case, a physical space has been designed so as to enable design-thinking-type activities on a very large scale – lecture theatre sized. This is great for conferences and for teaching. It means that more perspectives may be brought into the room and added to the crowdsourced mix. But, at the same time, we can all focus in on really small details. More often we use flat-floored open spaces, with plenty of wall space and reconfigurable furniture. The principles are the same. Design thinking cares for detail as much as it cares for the big picture of joined-up systems – all the varied elements that make up the things that we design today. Take any aspect of modern life, and focus in on the details that make it work well or not. For example, mobile phones. Small differences in the shape of the edges of a phone have an impact on users who hold and manipulate them with thousands of micro-actions over the years during which they are owned. Get that detail right and the design achieves what Chapman (2005) has called 'emotional durability' – we are more likely to keep hold of the phone for longer, rather than discarding it for the next latest model. Zoom out of that detail and consider how the phone works as part of multiple super-complex systems and networks of many kinds – electronic, human, material and so on. Design thinking is concerned at the same time both with the detail of the design-in-action, how that fits into ecosystems, communities and networks, and how we might change details so as to re-engineer the bigger picture for the benefit of real people and their worlds. That accounts for the *dynamism* needed to facilitate design thinking. Great facilitators are able to work the room, the mental models emerging in the minds of the participants, and the imaginative energy needed to design and redesign. Most importantly, they are able to quickly draw diverse people into this collective thinking and envisioning.

Diversity is essential, in developing a sufficiently rich picture of design challenges and the context for which we are designing, but also to give us realistic assessments of the ideas that emerge from collaboration. The current wave of interest in design thinking dates back to 2008 and a *Harvard Business Review* article by Tim Brown of the IDEO design consultancy (Brown, 2008). Brown described the benefits of creating three separate but connected spaces in which these activities may be hosted. Each space is adapted to a different purpose and a different kind of activity: the inspiration space, in which we collaboratively and freely build the rich picture of the world under consideration; the ideation space, in which we develop pro-totypes through which we 'build to think' (Kelley, 2001); and the implementation space, in which resources are available to implement designs that bubble up through the process. The spaces are intended to be open and continuously available. They are also ways of spreading design capabilities out into the community. Participating in design thinking is as much a learning experience for everyone as a way of solving problems. The central idea of Brown's 2009 book (Brown, 2009), which extends his 2008 paper, is that design innovation can revitalise organisations by transferring design capabilities from professional design teams to everyone across organisations that adopt its mentality and methods, and that this may be sustained without continual intervention. This understandably has proved to be a controversial claim among professional designers who have resisted the idea that they should give up some of their power, reject the figure of the genius designer, and instead trust in people and process. But as the systems we use become more malleable, more reconfigurable (especially digital systems), the notion that we are all designers, and should all be *designerly*, seems convincing. Design researchers have also made significant progress in describing what makes good designing work, such that we can more easily teach and support it – see especially Schön's (1987) *Educating the Reflective Practitioner*, Cross's (2007) *Designerly Ways of Knowing* and Lawson's (2005) *How Designer's Think*.

Design thinking at the University of Warwick

Design thinking itself has a history that pre-dates Brown's 2008 article, and a much broader interdisciplinary research base within academia. Buchanan's 1992 paper 'Wicked problems in design thinking' is one of the earliest uses of the term in its modern sense (Buchanan, 1992). Buchanan's claim was that professional designers have evolved concepts and tech-niques that can be used to address, or dissolve, even the most intractable 'wicked problems' (as originally defined by Rittel, 1972). These concepts have adapted ideas from many fields (but especially anthropology and psychology), following a needs-driven, pragmatic eclectic attitude. Design thinking today follows in this tradition, but seeks to take these powerful ideas and techniques out of the studio and to spread them among 'ordinary' people, as a form of social empowerment.

Thinking back to the experience of design thinking at Monash University in 2018, the question we asked ourselves at Warwick University was 'How do we develop great design thinking facilitators who are able to work this kind of magic for us and the wider world?' Our focus has always been on developing students as active participants in the design of higher education. But might it be that these capabilities are rare natural talents? Perhaps it takes someone really special to be able to 'work the room' effectively? If so, could we really hope to use the approach for wider social benefits?

Since 2018 we have established a small suite of design thinking modules for undergraduates and postgraduates. In 2021 we will run a 1-week-long version of our approach as part of the humanitarian engineering master's degree course. We think we have proved that anyone, from any background, can become good at doing it. Using a design thinking approach informed by the classic book *Design is Storytelling* (Lupton, 2017), Bo Kelestyn and I reframed our modules as a journey in which each week students go out into the world to work on 'design challenges' together. They then come back together to reflect on experiences, to consider design concepts and practices relevant to what they have encountered. Each session begins with a 'reflective jam' in which we practise the kind of design dialogue that moves between detail and big picture, designs-in-action and designs-as-imagined. For up to an hour, we sit in a circle and practise the reflective-designerly dialogue characteristic of designers. The jam is punctuated with concepts and examples, but always introduced in the context of their real-world application. This is important – designers have little time for theory for its own sake. Concepts are created as the exigencies of the design process require them, through reflection-on-action (as Donald Schön says). The reflective jam is followed by a series of exercises that deepen understanding, develop skills, provoke through challenges and exercise developing capabilities. We then lead into next-level design challenges, progressing from working in pairs towards working in bigger and more complicated teams. To begin with we focus on seeing the world through the eyes of a designer, and seeing them through the diverse perspectives of real people – with an emphasis on developing empathy as a technique. Our initial 'design anthropology' challenge sees the students going out into cafes to observe how they work (or in some cases not work) and how staff and customers work them. As we progress through the 10 weeks, we shift towards developing creative practices to generate and test fresh design ideas. Towards the middle of the module, we turn one of our weekly workshops into an authentic creative industries symposium, open to the public, in which successful design innovators come into the university to share their stories. In 2020 this was led by Catherine Allen, of the world-leading VR and immersive experience company Limina Immersive. This is a great way to make it real for the students.

The transformative journey of becoming designerly is presented as an adventure that we undertake together, developing our individual and collective design capabilities as a 'capability stack': knowing design concepts (at the base), participating in design thinking, facilitating design thinking (this is especially challenging), managing projects, and researching design thinking (with an emphasis on interdisciplinary dimensions, as our students come from many different disciplines). To help our students with the challenge of becoming facilitators, we use a workshop on risk-taking and empathy led by the brilliant physical theatre company Highly Sprung. Our students do find this all challenging, and some do struggle with this very different kind of learning experience. But the results are wonderful. And, in time, will have a great impact in the world beyond the university.

We are confident that we have been able to 'bottle' the design thinking approach, and structure a learning process for our students, so that they can work towards being great facilitators. And we can see this is having an impact for them. But the bigger challenge will be to take this and work with more communities, in our local area (Coventry, an economically challenged city in England), and throughout international connections (e.g. in Southern Africa). To achieve this much greater reach, we must explore how we can take our on-campus workshop methodology and make it work at a distance, online. That's a big challenge. But, using the design thinking approach, we are confident we can tackle it.

REFERENCES

All India Council for Technical Education (2021) Dashboard. https://facilities.aicte-india.org/dashboard/pages/dashboardaicte.php (accessed 12/04/2021).

Brown T (2008) Design thinking. *Harvard Business Review* **86(6)**: 84–92.

Brown T (2009) *Change by Design: How Design Thinking Transforms Organizations and Inspires Innovation*. HarperCollins, London, UK.

Brundage M (2016) Artificial intelligence and responsible innovation. In *Fundamental Issues of Artificial Intelligence* (Müller V. (ed.)). Springer, New York, NY, USA, pp. 541–552.

Buchanan R (1992) Wicked problems in design thinking. *Design Issues* **88(2)**: 5–21.

CBHA (Consortium of British Humanitarian Agencies) (2010) *Core Humanitarian Competency Framework*. CBHA, London, UK.

Chapman J (2005) *Emotionally Durable Design*. Taylor and Francis, London, UK.

CHS Alliance (2017a) *Core Humanitarian Competency Framework*. CHS Alliance, Geneva, Switzerland.

CHS Alliance (2017b) CHS Alliance (2017b) Get support: Core Humanitarian Competency Framework. https://www.chsalliance.org/get-support/article/launching-the-core-humanitarian-competency-framework-chcf-project/ (accessed 12/04/2021).

Cross N (2007) *Designerly Ways of Knowing*. Springer, New York, NY, USA.

Davis J and Lambert B (2002) *Engineering in Emergencies*, 2nd edn. Practical Action, Rugby, UK.

Devex (2021) *Devex newsletters*. https://pages.devex.com/newsletter-management.html (accessed 12/04/2021).

EFOD (2021) http://www.efod.org.uk (accessed 12/04/2021).

EWB (2021) http://ewb-international.com (accessed 12/04/2021).

GAUC (Global Alliance for Urban Crises) (2019) *The Urban Competency Framework*. http://urbancrises.org/resource-library/ (accessed 12/04/2021).

Gera I and Singh S (2019a) An Inquiry into the impact of the fourth industrial revolution on employment: a review. *Proceedings of 10th International Conference on Digital Strategies for Organizational Success, Gwalior, India*. https://ssrn.com/abstract = 3315169 (accessed 12/04/2021).

Gera I and Singh S (2019b) A critique of economic literature on technology and fourth industrial revolution: employment and the nature of jobs. *Indian Journal of Labour Economics* **62(4)**: 715–729, 10.1007/s41027-019-00191-8.

Government of India (2021) Welcome to National Service Scheme. https://nss.gov.in (accessed 12/04/2021).

ICE (Institution of Civil Engineers) (2019) *Continuing Professional Development Guidance*. ICE, London, UK.

Indian Department of Science and Technology (2020) S&T for socio economic programme. https://dst.gov.in/scientific-programmes/st-and-socio-economic-development (accessed 12/04/2021).

Indian MHRD (Ministry of Human Resource Development) (2010) *Annual Report 2009–10*. Ministry of Human Resource Development, Government of India, New Delhi, India. https://mhrd.gov.in/sites/upload_files/mhrd/files/document-reports/AR2009-10.pdf (accessed 12/04/2021).

Indian MHRD (2014) *Guidelines for Choice Based Credit System*. Indian MHRD, New Delhi, India. https://pib.gov.in/newsite/PrintRelease.aspx?relid=113370 (accessed 12/04/2021).

Indian MHRD (2015) *MOOCS: Massive Open Courses: An Initiative under National Mission on Education through Information & Communication Technology (NME-ICT) Programme*. Indian MHRD, New Delhi, India. https://www.aicte-india.org/downloads/MHRD%20moocs%20guidelines%20updated.pdf (accessed 12/04/2021).

Indian MHRD (2020) *Unnat Bharat Abhiyan 2.0: A Flagship program of Ministry of Human Resource Development*. Indian MHRD, New Delhi, India. https://unnatbharatabhiyan.gov.in/app/webroot/files/presentations/regional_workshop/UBA%20Regional%20Workshop%20-%20About%20UBA%2018-7-2018.pdf (accessed 12/04/2021).

Kelley T (2001) Prototyping is the shorthand of design. *Design Management Journal* **12(3)**: 35–42, 10.1111/j.1948-7169.2001.tb00551.x.

Kothari Commission (1966) *Report of the Education Commission: 1964–66*. Ministry of Education, Government of India, New Delhi, https://ia800206.us.archive.org/26/items/ReportOfTheEducation Commission1964-66D.S.KothariReport/48.Jp-ReportOfTheEducationCommission1964-66d.s.kothari.pdf (accessed 12/04/2021).

Lawson B (2005) *How Designers Think: The Design Process Demystified*. Routledge, London, UK.

Lupton E (2017) *Design is Storytelling*. Cooper Hewitt, Smithsonian Design Museum, New York, NY, USA.

Médecins Sans Frontières (2018) International Activity Report 2018: 2018 in figures. https://www.msf.org/international-activity-report-2018/figures (accessed 12/04/2021).

NAAC (National Assessment and Accreditation Council) (2019) *NAAC Institutional Accreditation: Manual for Self-study Report – Universities*. NAAC, Nagarbhavi, India. http://naac.gov.in/images/docs/Manuals/University-Manual-11th-January-2019.pdf (accessed 12/04/2021).

NITI Aayog (2018) NITI Aayog launches global hackathon on artificial intelligence. https://pib.gov.in/PressReleasePage.aspx?PRID=1555126 (accessed 12/04/2021).

ODI (2021) Subscribe to ODI updates. https://www.odi.org/newsletter-sign-up (accessed 07.02.2021).

O'Toole R (2015) *Transdisciplinary Studies of Design Thinking for the [Re]making of Higher Education*. PhD thesis, University of Warwick, Coventry, UK.

Oxfam India (2018) *India Inequality Report 2018: Widening Gaps*. Oxfam India, New Delhi, India. https://www.oxfamindia.org/sites/default/files/WideningGaps_IndiaInequalityReport2018.pdf (accessed 12/04/2021).

Ramachandra HD and Kumar A (2003) Engineering education in India. *Productivity* **44(2)**: 187–194.

RedR UK (2018) *Engineering in the Humanitarian Sector: Learning Needs and Challenges*. RedR UK, London, UK.

RedR UK (2021a) *Affiliate scheme: become an aid worker*. https://www.redr.org.uk/Get-Involved/Affiliate-Scheme-Become-an-aid-worker (accessed 12/04/2021).

RedR UK (2021b) *Red Alert magazine*. https://www.redr.org.uk/About/Red-Alert-Magazine (accessed 12/04/2021).

RedR UK (2021c) Training and learning for organisations and individuals. https://www.redr.org.uk/Training-Learning (accessed 12/04/2021).

Reed B and Fereday E (2019) Developing professional competencies for humanitarian engineers. *Proceedings of the Institution of Civil Engineers – Civil Engineering* **169(5)**: 49–56, 10.1680/jcien.15.00046.

Rittel H (1972) *On the Planning Crisis: Systems Analysis of the First and Second Generations*. Institute of Urban and Regional Development, Berkeley, CA, USA.

Schön DA (1987) *Educating the Reflective Practitioner*. Jossey-Bass, San Francisco, CA, USA.

Shaha S and Ghosh S (2012) Commissions and committees on technical education in independent India: an appraisal. *Indian Journal of History of Science* **47(1)**: 109–138.

Shelter Forum (2021) http://www.shelterforum.info (accessed 12/04/2021).

Singh S (2002) *Status of Social Science and Humanities component in Engineering Education: Findings from the Engineering Colleges of Haryana and Delhi*. National Technical Manpower Information System (NTMIS), Nodal Centre for Haryana and Delhi, Delhi College of Engineering, Delhi, India.

Singh S (2005) Social science component of engineering curricula in India. *Journal of Engineering Education* **18(3)**: 39–44.

Singh S (2013) Social responsibility of engineering institutions for inclusive growth. *Universities News* **51(3)**: 1–4, 12.

Singh S (2014) *Economics for Engineering Students*, 2nd edn. IK International Publisher, New Delhi, India.

Singh S (2016) Integrating social responsibility of university and corporate sector for inclusive growth in India. *Higher Education for the Future* **3(2)**: 183–196.

Singh S (2019) The fourth industrial revolution, women engineers and SDGs: an exploratory study with special reference to India. *International Conference on Women in Science, Technology, Engineering and Mathematics (STEM) (INWES), Kathmandu, Nepal*.

Sphere (2018) *The Sphere Handbook: Humanitarian Charter Minimum Standards in Humanitarian Response*, 2018 edn. Sphere, Geneva, Switzerland.

The New Humanitarian (2021) Sign up: newsletters and email alerts. http://www.thenewhumanitarian. org/subscribe (accessed 12/04/2021).

UGC (University Grants Commission) (2003) *Six Months Module Syllabus for Environmental Studies for Under Graduate Courses*. UGC, New Delhi, India. https://www.ugc.ac.in/pdfnews/ 2269552_environmentalstudies.pdf (accessed 12/04/2021).

UN (United Nations) (2019) *World Urbanization Prospects: The 2018 Revision*. UN, New York, NY, USA.

UN (2021) Sustainable Development Goals. https://sustainabledevelopment.un.org (accessed 12/04/2021).

UN Office for the Coordination of Humanitarian Affairs (2021) All jobs. https://reliefweb.int/jobs (accessed 12/04/2021).

UNESCO (UN Educational, Scientific and Cultural Organization) (2018) Harnessing artificial intelligence to advance knowledge societies and good governance. https://youtu.be/ 77LNQq9s3tU (accessed 12/04/2021).

University of Warwick (2021) Humanitarian engineering. https://warwick.ac.uk/fac/cross_fac/iatl/ study/humanitarianengineering/ (accessed 12/04/2021).

Humanitarian Civil Engineering: Practical solutions for an interdisciplinary approach

ice Publishing

Georgia Kremmyda
ISBN 978-0-7277-6468-3
https://doi.org/10.1680/hce.64683.163
ICE Publishing: All rights reserved

Index